THE WHICH? BOOK OF
WIRING & LIGHTING

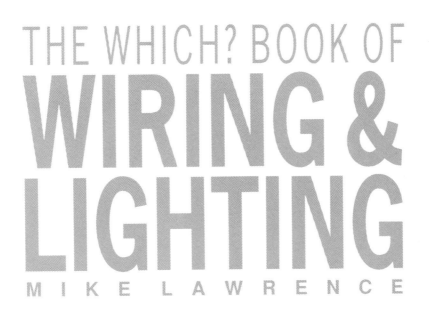

THE WHICH? BOOK OF
WIRING &
LIGHTING

MIKE LAWRENCE

Technical consultants
Trevor E. Marks A.R.T.C.S., C.Eng, M.I.E.E., M.Inst.M.C., F.Inst.D.
Anthony Byers

CONSUMERS' ASSOCIATION

Which? Books are commissioned and researched by The Association
for Consumer Research and published by Consumers' Association,
2 Marylebone Road, London NW1 4DF

First edition September 1993
Reprinted October 1994

Designed by Eve White
Cover photograph by John Parker
Photography by Dave Copsey
Illustrations by Peter Harper

Photograph on pages 16 and 17 reproduced by kind permission of
The Robert Harding Picture Library

Acknowledgements to Ashley & Rock Ltd, Crabtree Electric Industries
Ltd, Marbourn Ltd, MK Electric Ltd and Philips Lighting Ltd for their
generous assistance in supplying items for use in the photographs
throughout this book

British Library Cataloguing in Publication Data:
Lawrence, Mike
The Which? Book of Wiring and Lighting
I. Title
621.31924

ISBN 0-85202-551-3

Typeset by Delta Print, Watford, Hertfordshire
Printed and bound in Great Britain by Scotprint Ltd., Musselburgh

CONTENTS

INTRODUCTION

The electricity supply is the one essential element of the home that seems to strike fear into the hearts of householders everywhere. People who have built home extensions, knocked down walls and installed their own central heating without a qualm simply shy away from doing their own wiring work, on the very reasonable grounds that electricity can kill and so is best left alone.

Of course electricity is dangerous, but so are many other do-it-yourself activities if they are carried out without due care and attention; far more people are injured or killed every year in falls from ladders than from doing their own wiring. All you need to work safely with electricity is a clear understanding of how it behaves coupled with the ability and patience to approach any job methodically.

You need no special skills to do your own wiring work – nothing like those needed for cutting complex woodworking joints or soldering plumbing fittings – and hardly any specialist tools. The materials are widely available from retail outlets and do-it-yourself superstores; you no longer have to queue up with the trade at local contractors' or wholesalers' counters to buy your supplies (although you may save money by doing so).

The one thing you do need is knowledge, and that is what *The Which? Book of Wiring & Lighting* aims to provide. It starts by introducing the basic tools, equipment, materials and fittings you will need to do your own wiring work. It then explains in detail how electricity reaches your home, how it is distributed to the various individual circuits that power the lights and appliances you use, and how each of these is connected to the power supply. It also stresses the vital importance of the various protective devices built into every wiring system.

Succeeding sections then deal with general wiring techniques and a wide range of individual wiring projects, from moving or adding light fittings and socket outlets to providing complete new circuits, taking power out of doors or completely rewiring an old system. Finally, there is an invaluable section on tracing and repairing electrical faults.

the Wiring Regulations

The legal situation regarding do-it-yourself electrical work in the United Kingdom is unusual. In many countries licensed electricians must carry out – or at least supervise – the work; here, except in Scotland (where electrical systems are covered by the Scottish Building Regulations and any work done **must** conform with their requirements), anyone can do his or her own wiring work without any legal restriction or any statutory guidelines to follow. The only potential restraint lies with local electricity companies, who have the right to test any wiring work and to refuse to supply an unsafe installation.

In the absence of any legal guidelines, any work you carry out should instead comply with the Regulations for Electrical Installations, drawn up by the Institution of Electrical Engineers (IEE) and used by all professional electricians. They are generally known as the IEE Wiring Regulations. These Regulations are revised from time to time; the current edition (the 16th) has been in force since 1991 and was made into a British Standard (BS7671) in 1992.

All the instructions in The Which? Book of Wiring & Lighting comply with the Regulations, but if you feel unsure of your ability to carry them out to the letter, always use a qualified electrician – a member of either the Electrical Contractors' Association (ECA) or on the roll of the National Inspection Council for Electrical Installation Contracting (NICE-IC). You can get names of qualified people working in your area from these two bodies or from your local electricity company showroom.

a Word Of Warning

All the instructions given in The Which? Book of Wiring & Lighting assume that your home's wiring system is safe and has been installed to meet the requirements of the edition of the Wiring Regulations applicable at the time of its installation or subsequent modification. All the specifications for equipment assume you have a 240V supply. Cable sizes given in this book are correct for the majority of domestic installations; however, you should check with a qualified electrician when installing equipment with a high current demand or if you are running cables over a long distance, together in a confined space, through insulation or in a high-temperature environment.

If you have any reason to suspect that your system may have been extended or modified in an unorthodox manner, or if there is evidence that rubber-covered cables are still in use anywhere on the system, have it inspected by a qualified electrician before attempting to carry out any of the jobs described here. Rubber insulation was used until the 1940s; it becomes brittle with age and crumbles away if disturbed.

Above all, do not tackle any electrical work unless you understand completely how to carry it out and are confident of your technical ability to do so correctly. 'He did it himself' is a poor epitaph for the careless or incompetent do-it-yourself electrician.

PROBLEM CHECKLIST

The section on tracing and repairing faults on pages 134-42 gives full details of how to replace fuses, flex, cable and wiring accessories, but if you are aware only of a symptom, this brief checklist will help you track down the cause of the problem and tell you where to find the solution.

A LIGHT DOESN'T WORK

1 Check whether the light bulb (correctly called the lamp) has failed. Turn off the switch, remove the lamp and replace it with a new one of the appropriate type and wattage.

2 With ceiling or wall lights, turn on other lights to check whether the fuse or miniature circuit breaker (MCB) protecting the lighting circuit has cut off the supply. This could have been caused by a short circuit, due perhaps to a loose connection or a broken wire, or to overloading of the circuit caused by fitting high-wattage lamps. Correct the fault (pages 135 and 137) and replace the fuse or reset the MCB (page 136).

3 With plug-in table or standard lamps, check the bulb first (see Step 1). If it still does not work, unplug the lamp and check the connections within the plug, whether the plug fuse has blown, the connections within the lamp and the flex continuity. Remake loose connections (page 137), replace faulty flex (page 137) or fit a new 3-amp fuse (page 136) as required.

AN APPLIANCE DOESN'T WORK

1 If there is no sign or smell of overheating or burning, unplug the appliance and plug something else into the socket. If this works, the fault lies within the original appliance. If it does not, there is probably a circuit fault. See *A circuit is dead* for what to do next.

2 If the appliance feels hot or is smoking, do not try to use it; unplug it and have it examined by a service engineer. If not, unplug the appliance, open its plug, check the flex connections at the terminals and test the fuse as for plug-in lamps (pages 136 and 137).

3 With portable appliances (still unplugged), try to gain access to the terminal block where the incoming flex is connected. Check that each core is securely linked to its terminal, and inspect the flex for signs of damage or burning; then test the flex continuity (page 137). Replace any cover plates or fully reassemble the casing before plugging the appliance in again. If it still fails to work, take it to a service engineer for repair.

A CIRCUIT IS DEAD

1 If none of the lights or appliances on a circuit works, switch off all the lights and disconnect the appliances. Then check whether the circuit fuse or MCB has cut off the supply, and replace/reset it if it has (page 136).

2 If the fuse blows again or the MCB trips to off immediately, the fault is on the circuit wiring. With the mains power off, open up wiring accessories to check for loose connections or short circuits (pages 139-40). Repair the fault and replace the fuse/reset the MCB.

3 If it does not, switch lights on or plug in appliances one by one. If a particular light or appliance blows the circuit fuse or trips the MCB, isolate it for testing. Check too whether that light or appliance is overloading the circuit although itself functioning normally. NEVER fit a fuse with a higher rating to try to prevent persistent fuse failure.

4 If the circuit protective device still operates, the fault is on the circuit wiring. If you have drilled or nailed into a cable, turn off the power and repair the damage (page 139). Otherwise, call in a professional electrician to test the circuit and trace the fault.

THE WHOLE SYSTEM IS DEAD

1 If you appear to have a total power failure check to see whether neighbours also have a power cut. Remember that a single-phase fault on the three-phase system used to supply domestic premises will black out only those houses connected to that phase. Report any power supply interruption to your local electricity company's 24-hour emergency number (listed under Electricity in your phone directory). The company will be able to confirm if there is a fault, and also to tell you when power is likely to be restored.

2 If yours is the only house without power and your system is protected by a whole-house residual current device (RCD), check whether it has tripped to off. If it has, reset it.

3 If the RCD will not reset, the fault is still present and you should run through the appliance and circuit checks outlined above to track it down. Alternatively, call in an electrician. If your RCD keeps tripping off for no apparent reason – so-called nuisance tripping – report this to your electricity company.

4 If you do not have an RCD and yours is the only house without power, the fault lies either in your incoming supply cable or in your main service fuse. Call the 24-hour emergency number and ask for an engineer to come to correct the fault. This service is free of charge.

SAFETY FIRST

Before you carry out any electrical work, make sure that you are following all these safety guidelines. If you do not, your safety – and that of your family – will be at risk.

- Do not carry out any electrical work unless you are confident that you know what you are doing and are able to complete the job correctly and safely. Be especially careful if you suffer from colour blindness.

- Always turn off the power supply at the main isolator switch or RCD before working on your home's wiring system. If you are working on just one circuit, remove the circuit fuse or switch off the MCB before restoring power to the other circuits. Take care when working on two-way switching arrangements, which can (but shouldn't) link two lighting circuits. Always unplug appliances before trying to inspect or repair them.

- Always double-check all connections inside plugs, appliances and wiring accessories, to ensure that flex and cable cores are linked to the right terminals and that they are securely fixed.

- Never omit the earth connection. The only circumstances where one is not needed is in the flex run to non-metallic lampholders and double-insulated power tools or appliances.

- Never touch fittings or appliances with wet hands, or use electrical equipment out of doors in wet conditions. Always provide RCD protection for outdoor appliances.

Shock treatment

If an appliance or a wiring accessory gives you a minor shock, stop using it at once. Have the former checked by an electrical appliance repair expert for earth safety, and replace the latter if it is damaged and live parts are exposed.

If someone receives a major shock, try to turn off the source of the current as fast as you can. If you cannot, grab clothing (NOT bare flesh, or you will get a shock too if the power is still on) and drag the victim away from the power source.

Lay the victim flat on the back with legs slightly raised if conscious but visibly shocked, turn the head to the side to keep the airway clear and cover with a blanket. Do not give anything to drink or smoke. Flood burns with cold water, then cover them with a clean sterile dressing; don't apply ointments or remove loose skin. Call an ambulance.

Lay the victim in the recovery position if unconscious. Keep the airway clear by tilting the head back and bringing the jaw forward. Cover with a blanket and call an ambulance immediately. Then monitor breathing or heartbeat continuously; if either stops, give artificial ventilation or external chest compression as necessary until the ambulance arrives.

WIRING BASICS

THE ELECTRICITY SUPPLY

To understand how electricity works, it may be helpful to compare it with man's far older domestic and industrial ally, water. What makes water flow through a pipe or other conduit is a pressure difference between inlet and outlet. This can be provided by the force of gravity, for example when water is released from a reservoir by opening the sluice gates in the dam, or else by a pump. The greater the pressure difference between the ends of a pipe, the greater the rate at which the current of water flows through it.

On its way between two points, water can also do work – for example, by flowing over a water wheel which in turn can drive a machine. In a domestic heating system a pump circulates water through a heat exchanger, where it absorbs heat and then circulates it to radiators around the house, again doing work by delivering heat to the rooms.

Electricity works in a similar way. It too needs a circuit to flow round, and is driven round it by a pressure *(potential)* difference between the inlet and outlet ends of the circuit. This potential difference is measured in units called **volts** (V for short), and in the UK the potential difference delivering mains electricity to our homes is usually about 240 volts. The rate of flow of electricity – the current – is measured in **amperes** (A for short). As electricity flows round a circuit it too can do work – for example, heating up a lamp filament or a fire element, or driving an electric motor. The work it does is measured in **watts** (W for short). For most practical purposes the work equals the voltage multiplied by the current that an appliance draws (watts = volts x amps).

Water will clearly flow more easily through a wide, smooth-bore pipe than through a narrow, rough one which resists the water flow. The same principle applies in electricity, where the term *impedance* is used to describe whether electricity flows freely, poorly or not at all through a particular material. Conductors – mainly metals – have low impedance to the flow of electricity and so are used to carry it round circuits to where it is needed, while insulators – mainly plastics, ceramics and rubber – have high impedance and are used to prevent electricity from 'leaking' from the conductor carrying it. Impedance is measured in units

called **ohms** (Ω for short); the higher the impedance, the lower the current flow.

Conductors have to be an adequate size for carrying the current of the circuit without generating too much heat or impeding the flow to the extent that the voltage at the exit from the conductor is much lower than at its entrance. In practice this means that the higher the current and, to some extent, the longer the circuit, the thicker the conductors (the circuit cables) have to be.

A plumber refers to 'flow' and 'return' pipes when describing a central heating system; an electrician uses the terms 'live' (or phase) and 'neutral' to describe the flow and return parts of an electrical circuit in much the same way. This terminology does **not** imply that any electrical conductor or component described as neutral is 'safe'; **all** parts of an electrical circuit carry electricity, just as all parts of a plumbing circuit contain water.

The electricity supply reaches individual properties via an underground or overhead service cable. Shortly after entering the house it terminates in a sealed box called the **service head** or **cut-out**. This contains the main **service fuse**, which is present to prevent the property from demanding more current than the supply cable can safely carry without overheating, and which will 'blow' if its current rating is exceeded. Most modern homes have an 80- or 100-amp supply to cope with today's increasing demand for electricity, but older homes may have only a 60-amp supply and a correspondingly smaller service fuse.

From the cut-out two thick single-core cables, one with a red sheath and one black, run to your electricity meter. This records the amount of electricity the system uses in units called kilowatt-hours (kWh for short), and may have dials or a digital counter. Up to this point, all the equipment is the property of the local electricity supply company, and it is an offence to tamper with it in any way.

If the property has been wired to make use of cheap night-rate electricity, there will be a dual-rate meter and a time switch wired in after the cut-out instead of a conventional meter. This arrangement switches the meter to record the consumption of electricity at the two different charge rates at pre-set times.

From the meter, two more single-core cables called **meter tails** run to the home's power distribution centre, known colloquially as the fusebox. In modern homes, this is a one-piece enclosure called a **consumer unit** which contains the system's main isolating switch, and from where electricity is distributed to the various circuits in the building – to lights, socket outlets and appliances such as cookers and electric showers. If the property uses cheap-rate electricity, there will be a second consumer unit (often contained within the same enclosure) wired up to supply the circuits or appliances concerned – chiefly storage and immersion heaters. Economy 7 is the most widely used off-peak tariff. However, check with your local electricity supplier, since there may be more advantageous tariffs available.

Older homes often have wiring systems that have been extended over the years, with several small fuseboxes supplying different circuits and linked by a spaghetti-like maze of cables, often controlled by a separate main on-off switch. Such arrangements are generally long overdue for replacement.

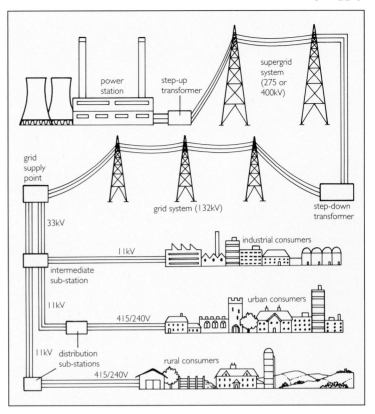

Above: Electricity generated at power stations is transmitted at high voltage across the country. Transformers step the voltage down at various stages, eventually to the 240-volt supply used in the home

Right: Within the home, the incoming power supply is taken via the cut-out (which contains the main service fuse) to the electricity meter and then to the consumer unit. This distributes the supply to the various circuits in the house

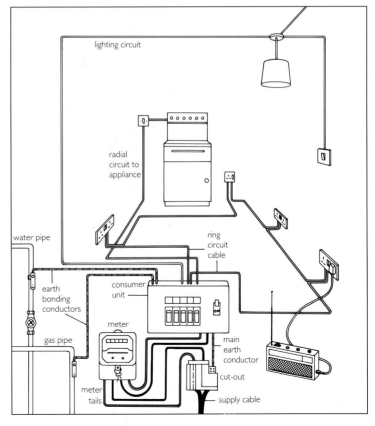

EARTHING PROVISIONS

One of the most important parts of any wiring system is the provision made for earthing it. In normal operating conditions electricity should flow rather like water round a plumbing system, with no leaks. When a leak does occur, a means must be provided for coping with the resulting overflow – an **earth-fault current** – before it can cause any damage, either to the system itself or to its users. The solution is to allow it to pass 'to earth'.

Every domestic wiring installation will be earthed in one way or another, usually by the local electricity supply company. Modern installations use what is known as **protective multiple earthing** (PME). The system is earthed via the supply cable's neutral cónductor, which is connected to the 'star point' of the local supply transformer; this is itself earthed at the sub-station. Within the property itself, a connection is made from the cut-out to a main earthing terminal, from where a separate **earth continuity conductor** (ECC) is run round the house to link all the various components of the system so that an earth fault current can flow safely to earth at any point. This sort of installation is known as a *TN-C-S* system.

Older homes frequently make use of the metallic sheath on the supply cable as the main earthing provision; this is known as the *TN-S* system of earthing. TN-S systems are being gradually replaced by TN-C-S systems.

Houses in rural areas or with overhead power supplies may have no earth connection via the supply cable; instead, earthing is provided by means of an earth electrode driven into the ground beneath or beside the house; this is known as *TT* earthing. As with PME earthing, an earth continuity conductor then links the various components of the house wiring system to the main earth terminal. Effective earthing can be difficult to achieve with *TT* earthing, and circuits supplying socket outlets must therefore be given additional protection by the use of a residual current device (see right).

Because of the importance of having satisfactory earthing arrangements, it is essential that earth connections should never be tampered with. They should all carry a permanent stamped metal label warning against accidental or deliberate disconnection.

USER AND CIRCUIT PROTECTION

The purpose of the consumer unit is not only to distribute electricity to the various circuits in the house but also, most importantly, to protect the system and its users from the dangers of shock and fires caused by electrical faults.

An electric shock is experienced when someone touches a live part of the system and current passes to earth through the body; the higher the current, and the longer the time for which it is received, the greater the risk of severe injury or death.

A fire may be caused in one of two ways. The first is by sustained overloading of any part of the system, causing insulation to fail and heat to ignite combustible materials. The second is by short-circuiting, where current-carrying conductors touch and form an arc between them, again possibly starting a fire.

The modern consumer unit contains two types of protective device to guard against shock and fire. The first is the **residual current device** (RCD), which detects earth-fault currents the instant they occur and isolates the supply in a fraction of a second. RCDs can protect all or part of the wiring system, and units with sensitivities ranging from 10 to 100mA (milliamps) are available for different situations.

The second protective device is the **circuit fuse** or **miniature circuit breaker** (MCB); each circuit has its own fuse or MCB to guard against overloading of that particular circuit.

MCBs are superior to fuses for a number of reasons. They are convenient to use to isolate a circuit for maintenance or extension work, and give a clear indication of which circuit is affected when they detect a fault and trip to off. They also have the advantage over fuses that they are not easily abused.

Fuses and MCBs also provide an element of protection from shock resulting from contact with something metallic that has accidentally become live – an appliance, for example – provided that the resistance of the current path to earth via the earth continuity conductor is low enough. However, an RCD is essential to provide protection against electrocution; it will not stop you getting an electric shock resulting from direct contact with live parts of the wiring system itself, such as conductors and terminals, but will cut the power off fast enough to stop it from killing you. New wiring installations must have RCD protection.

DISTRIBUTION EQUIPMENT

The electrical requirements of a modern home are a far cry from those of the '60s and '70s, when a couple of lighting and power circuits plus a supply to a cooker or immersion heater were regarded as generous. Nowadays a typical home may have three lighting circuits (two indoors, one outdoors), three power circuits, separate circuits to a cooker, a shower and an immersion heater, plus additional circuits controlling such things as mains-powered door bells, smoke detectors and burglar alarms, or themselves under timer control. There may be a separate circuit supplying an outbuilding, and if electric storage heaters are being used there will be a separate consumer unit supplying their circuits.

The consumer unit will be sized to match these circuit requirements, and also to provide the chosen level of earth-fault protection. It may also contain one or two spare ways, to allow for possible future expansion of the wiring system.

Electricity meters may be the old-fashioned type with clock dials (left) or the more modern type with a digital display, which is easier to read

METERS AND METER BOXES

In older homes the electricity meter is often sited in a highly inconvenient place – under the stairs or in a cupboard. In newer homes it is commonly housed in a special meter enclosure let into the house wall and accessible to the householder and electricity-company meter readers via a lockable door.

It is a good idea to read the meter whenever an electricity bill arrives, to check whether it has been read correctly or whether an estimated reading is reasonably accurate. Modern digital meters are straightforward to read, but older dial-type meters can be confusing.

To read one, start with the left-hand dial, which records units used in 10,000s of kilowatt-hours (kWh), and work across to the right, recording the reading from the dials marked 1,000, 100, 10 and finally 1 kWh. Ignore the red dial. Record the number the pointer has just passed on each dial; if it is between 3 and 4, for example, write down 3. If the pointer is directly on a number (5, say), check the dial on its right. If its pointer is between 0 and 1, write down 5 as the reading for the previous dial. If it is between 9 and 0, write down 4 for the previous dial: its pointer has not quite reached the 5. The reading on both the meters shown here is 26951.

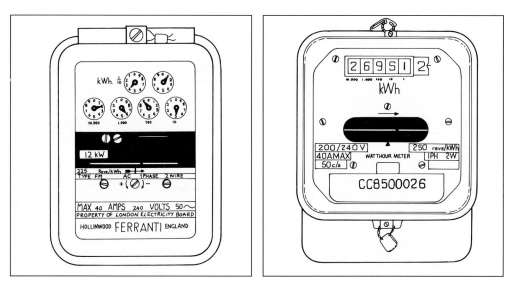

THE WIRING TOOLKIT

If you do even the most straightforward d-i-y work, you will find that many of the tools used for wiring are already in your toolkit: things like drills, hammers, screwdrivers and saws for example. However, there is a small range of specialist tools that you will need for jobs such as cutting and stripping flex and cable, and also for testing your work as you proceed. The A–Z list on pages 18–20 details everything you are likely to need for any job, from making con-

nections to routing cable around the house and installing fittings.

Always buy the best-quality tools you can afford and look after them with care to ensure that they give you good service and performance. There will be occasions when you can improvise rather than buy a tool you may use only once; alternatively, you may decide to hire rather than buy. You will find this advice given where it is relevant.

Lastly, even if you have a well-stocked toolkit, it is often difficult to find the tools and materials you need quickly in an emergency, particularly if the lights have failed. It is therefore a good idea to assemble a simple selection of tools and sundries, to place them in a suitable container and to keep them handy somewhere in the house so you can locate them easily when they are needed. See *Emergency toolkit* on page 19.

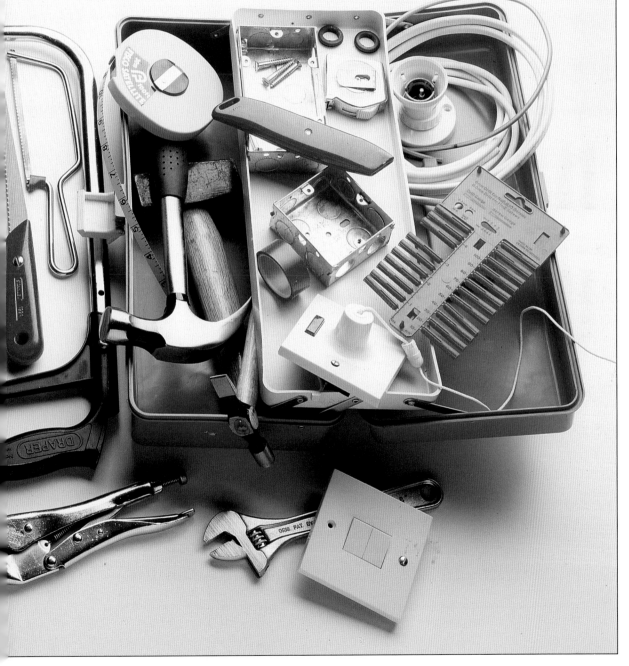

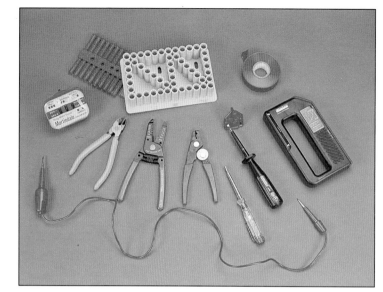

Specialist tools and equipment. Top row, left to right: socket tester, drilling guide, red PVC insulating tape; middle row: side cutters, wire strippers (two types), mains testers (two types), cable detector; bottom row: continuity tester

A–Z OF TOOLS

Brace and bits

A standard carpenter's brace is actually of little use for wiring work because of the clearance needed to sweep the handle round, but its smaller cousin, the **joist brace**, is ideal for drilling holes in confined spaces through floor joists and wall studs. The chuck is turned by moving the short lever back and forth while applying hand pressure to the domed head. It will accept all standard square-shanked bits.

Alternative: use an electric drill with a right-angle adaptor.

Brick bolster

You will need a sharp brick bolster 2⅜" wide by ⅛" thick for chopping out recesses in solid walls to take flush mounting boxes. It can also be used for cutting chases – channels for cable – in walls, and at a pinch for prising up floorboards.

Alternative: use a drill and cold chisel for cutting recesses, and a hired chasing machine for cutting chases.

Cable detector

This battery-powered device – essentially a small metal detector – is useful for locating the presence of cable and pipe runs hidden in walls or beneath floors. Some types can also indicate the positions of wall studs and ceiling joists by detecting the presence of the nails securing the plasterboard to the supporting timbers behind it. Hire one unless you have other uses for it.

Chasing machine

This power tool is designed specifically for cutting neat, parallel-sided chases in solid walls, ready to accept buried cable runs. Since you are unlikely to need it very often, it is best to hire one. You can also buy smaller **chasing attachments** for electric drills. Both create a lot of dust and can be very noisy when used in confined spaces, so wear safety goggles and a face mask, plus ear defenders if necessary.

Alternative: a brick bolster and club hammer.

Chisels

You will need a range of **wood chisels** for cutting notches in joists and wall studs. **Cold chisels** in various sizes will come in handy for neatening up recesses and wall chases, and for cutting holes through walls. A **brick bolster** is useful for prising up floorboards, or a crowbar could be used instead.

Continuity tester

This is a battery-powered device which is used to check that there is a continuous electrical pathway when you are unable to see this visually – along the cores of a length of flex, for example. It can also test cartridge fuses. The simplest types have two probes and an indicator light to show continuity; **test meters** can also act as continuity testers.

Drills

An electric drill with a selection of twist drills, wood bits and masonry drills is essential for making fixings and for drilling holes for cables to pass through. Cordless drills are particularly handy for when the power is off. Useful drill attachments include a right-angle adaptor, a chasing attachment and screwdriver bits. A plastic **drilling guide** is useful for preparing masonry walls before chopping out recesses to install flush mounting boxes.

Alternatives: a hand drill or joist brace.

Ear protectors

These give useful protection when using noisy tools, particularly when chasing walls or cutting holes in confined spaces.

Extension reel

You will need an extension lead to provide power for a light and to drive power tools (unless yours are cordless) when working in areas without a power supply. Always uncoil the lead fully before using it.

Face mask

A face mask is essential to protect your lungs when drilling, cutting or chasing, or at any time when dust will be in the air. A cheap version has a disposable face cup with an elastic band to hold it in place.

Filler

You will have a certain amount of repair work to do after fitting recessed mounting boxes and cutting chases in solid walls. Use **plaster** or a **cellulose filler**, applied with a filling knife or small pointing trowel.

Fixings

You will need a range of everyday fixing devices – screws, nails, wallplugs and cavity fixing devices – to secure things like mounting boxes to walls and light fittings to ceilings.

Hammers

You will need a **claw hammer** for all sorts of jobs, from driving nails and wood chisels to tapping mounting boxes and recalcitrant wallplugs into place. You will need a **club hammer** to drive a brick bolster or cold chisel when cutting recesses and chases in walls. Buy one with a moulded-on plastic hand guard if you do not own one already. For lighter work, such as fixing clips, use a 12oz **pin hammer**.

Hand lamp

You will probably need a movable mains-powered light source. Choose a hand lamp with a metal guard to protect the lamp and yourself. The guard also helps stop the lamp causing a fire if it is left near something combustible.

Immersion heater spanner

This is an example of a tool that does just one job – screwing the large hexagonal head of an immersion heater into a hot water cylinder. However, they are cheap to buy, so it is not worth hiring one.

Knife

A sharp handyman's knife is invaluable for the huge range of cutting jobs in wiring work. Make sure you have a supply of spare blades.

Mastic

You will need silicone mastic for sealing connections made out of doors – for example, round the baseplates of outside lights. Either buy a small hand-operated cartridge, or a larger one that fits a standard mastic gun.

---

emergency toolkit

Keep the following tools and sundries somewhere handy inside the house, in a container such as a plastic lunch box, so you can find them easily when all the lights go out:

- *Small terminal screwdriver*
- *Medium-sized screwdriver*
- *Pair of pliers*
- *Handyman's knife*
- *Torch plus a spare set of batteries*
- *Roll of PVC insulating tape*
- *Rewirable fuseholders already wired up with 5-, 15- and 30-amp fuse wire or replacement cartridge fuses to match the circuit ratings in your fusebox*
- *Replacement 3- and 13-amp plug fuses.*

Rewirable fuseholders cost very little, and having replacements already wired up means you can slot in the appropriate fuseholder to restore the power instead of having to fiddle about in the dark with fuse wire and a torch. If your consumer unit has spare capacity it is worth installing a light on its own circuit nearby so you can see to replace fuses or switch MCBs when the main lighting circuit goes.

Measuring tape

A measuring tape is essential for measuring everything from the positions of wiring accessories to the length of cable runs. It doesn't matter whether it measures in imperial or metric units; use whichever you are more comfortable with.

Neon tester

This resembles a small terminal screwdriver, but has a neon indicator built into the handle. This lights up when the cap on top of the handle is depressed if a component touched by the blade is live. You should never, of course, be working on anything that's live.

Pliers

These are useful for bending and twisting cable cores together before they are connected to the terminals of a wiring accessory. Types with insulated handles are available, but since you should not be working on live parts anyway, regard this as only an additional safety feature. A pair of **long-nosed pliers** is useful for tucking cores into relatively inaccessible terminals.

Plumb bob and line

A plumb line is handy for checking that cable runs are vertical, and it can also be used to fish cable down from above through the void in a stud partition wall.

Alternative: small bolt tied to string.

Safety glasses

Eye goggles are essential when you are cutting chases or doing any job during which timber, stone, plaster or metal particles are liable to fly into your face.

Saws

A specialist **floorboard saw** has a curved tip and teeth on the top edge of the blade too, allowing it to cut right up to skirting boards without the handle fouling the wall. It is used to cut across floorboards and also to sever the tongues of tongued-and-grooved boards that have to be lifted so cables can be run beneath them. A **tenon saw** or a **circular saw** can be used instead. A **jigsaw** is useful for cutting the tongues off tongued-and-grooved floorboards. A **padsaw** is invaluable for making cut-outs for mounting boxes in plasterboard walls and ceilings, and a **junior hacksaw** is essential for cutting PVC conduit and protective cable channel neatly to length.

Screwdrivers

You will need all the screwdrivers you can lay your hands on, in a range of sizes, for everything from tightening terminal screws to fixing mounting boxes and attaching accessory faceplates. The bare minimum is a thin-bladed **terminal screwdriver** for small terminal screws, a **medium-sized driver** for faceplate screws (which are always the slotted-head variety), and a **large driver** for fixing mounting boxes and the like. You will find a set of **cross-point drivers**, used with cross-head screws, a boon since you can sit the screw on the blade tip and offer it up to the fixing in confined spaces.

Alternative: a cordless screwdriver with interchangeable bits if you have a lot of fixings to make.

Side cutters

These are powerful cutters used for cutting flex and cable to length; nothing does it better, although you can use the cutting jaws in a pair of combination pliers as an alternative. As with pliers, insulated handles are only for additional safety.

Socket tester

This is a special plug-in tester which is used to check that socket outlets have been wired up correctly.

Alternative: careful visual inspection.

Spanners

Some appliances and light fittings have nut-and-bolt connections, so small and medium-sized adjustable spanners will be useful.

Spirit level

A small spirit level is useful for ensuring that mounting boxes, accessory faceplates and runs of trunking are level.

Tape

Red tape must be used to mark black switch cable cores to show that they are live. **PVC insulating tape** is also useful for making emergency temporary repairs to damaged flex, plugs and wiring accessories.

Test equipment

Professional electricians use a range of specialist test equipment which is not readily available to the do-it-yourselfer.

Apart from the **continuity tester** and **socket tester** mentioned earlier, you should rely on good practice to ensure that your wiring is up to standard and have the finished job inspected by a professional electrician. Rewiring or major extensions to your wiring system should always be tested by a professional electrician before being commissioned.

Torch

A torch is essential both for coping with unexpected blackouts and for peering into the dark recesses of underfloor voids and lofts. Make sure you have spare batteries available. A small version of the miner's lamp, worn on an adjustable headband, can be useful for work in confined spaces since it leaves both hands free.

Wire

Some galvanised or plastic-coated wire – the sort you use in the garden for training plants – will come in handy for fishing lengths of cable across floor and ceiling voids.

Wire strippers

These are designed to remove the insulation from cable and flex cores without damaging the conductors – vital with flex, where carelessly chopping through the fine strands can reduce the flex's current-carrying capacity and lead to overheating. The jaws can be adjusted to suit the core diameter.

BUYING ELECTRICAL GOODS

The amount of wiring work you plan to do will affect where you buy the tools, equipment and fittings you need. For a job requiring just an accessory or two and a metre of cable, it probably matters little whether you visit the local hardware shop or a d-i-y superstore. However, for larger-scale projects it is well worth shopping around. You have several options.

D-I-Y SUPERSTORES

The obvious first choice. Most stock a reasonable selection of tools and equipment, often with a choice of plain white or brass wiring accessories, which may be own-branded. Cable and flex are sold in complete drums and by the metre. You will also find small consumer units complete with fuses or MCBs, a range of light fittings and ancillary items such as burglar alarms, door bells and telephone extension kits. Don't expect much in the way of technical advice, though the packaging may carry useful instructions.

HARDWARE SHOPS

These tend to stock a limited range of the most popular accessories – light switches, socket outlets and ceiling roses – plus the commonest types of cable and flex, sold by the metre only. They may not have the specialist tools you need. Ideal for small projects where convenience is more important than price.

ELECTRICITY COMPANY SHOPS

These offer a range of the most widely-used accessories, lamps, cable and flex, but not the more unusual items, and they seldom stock tools. They will also supply names of qualified electricians working in your area if you need one, and will arrange tests on your completed work (but obtain a quote first).

SPECIALIST RETAILERS

Electrical retail outlets are becoming more widespread in high streets, and offer the combination of excellent stocks of tools, equipment and accessories plus sound technical advice if you need it. Lighting specialists offer the widest choice of light fittings, although not necessarily at the lowest prices.

British Standard products

Wherever you decide to shop, buy only accessories that conform to the relevant British Standards. This will ensure that you choose products which are well made and suitable for their stated purpose. The BS number should be printed on the packaging or stamped on the accessory itself.

- *Cable BS6004 (marked on drum and possibly on sheathing)*
- *Ceiling roses BS67*
- *Ceiling switches BS3676*
- *Circuit cartridge fuses BS1361*
- *Coaxial socket outlets BS3041*
- *Consumer units BS5486*
- *Cooker control units BS4177*
- *Dimmer switches for tungsten filament lighting BS5518*
- *Double-pole (DP) switches BS3676*
- *Flex BS6500 (marked on drum and possibly on sheathing)*
- *Fused connection units (FCUs) BS5733*
- *Junction boxes BS6220*
- *Lampholders (bayonet cap) BS5042*
- *Lampholders (Edison screw) BS6776*
- *Luminaire support couplers (LSCs) BS7001*
- *Metal mounting boxes BS4662*
- *Miniature circuit breakers (MCBs) BS3871*
- *Passive infra-red sensors BS4737*
- *Plastic mounting boxes BS5733*
- *Plateswitches (light switches) BS3676*
- *Plug fuses BS1362*
- *Plugs BS1363 (BS546 for round-pin types)*
- *Residual current devices (RCDs) BS4293*
- *Shaver sockets (for rooms other than bath/shower rooms) BS4573*
- *Shaver supply units (and other transformers) BS3535*
- *Socket outlets BS1363 (BS546 for round-pin types)*

ELECTRICAL WHOLESALERS

These stock the widest range of wiring accessories of all types, often from a number of different manufacturers, and will also stock specialist tools. However, you need to know exactly what you want before you approach them. Contact them for estimates if you are planning large-scale wiring projects; they are generally the cheapest for big orders.

WIRING MATERIALS

A household wiring system is made up of a fairly small range of components, all designed for a specific job. Make sure that all the wiring materials you buy are made to the relevant British Standard (see page 21).

CABLE AND FLEX

It is important to understand clearly the distinction between cable and flex (which is short for flexible cord). Cable is used for all fixed wiring work on the various circuits around the house, while flex is used to connect electrical appliances and pendant lights to an accessory on the mains supply. The only exception to this is the final connection to free-standing electric cookers, which is made with cable because flex is not made in a large enough size for the job.

CABLE

Cable has a flattened oval cross-section, and is relatively stiff. The current-carrying conductors are contained within a tough PVC outer sheath. This is usually grey or white; the latter is preferred for surface-mounted wiring.

You will be using cable with three conductors (called **cores**) for almost all your wiring work. Two of these have colour-coded PVC insulation: **red** cores are used for **live** (phase) connections and **black** cores for **neutral** ones. The third core is bare and lies between the two insulated cores. This is used as the **earth continuity conductor** (ECC) to provide a continuous path right round the system along which current can flow harmlessly to earth if a fault should develop (see page 66 for more details). When the core is exposed at a connection to an earth terminal at any point on the circuit, it must be covered with a length of green-and-yellow PVC sleeving. This type of cable is commonly called **twin-and-earth**.

Cable with four conductors – three insulated, one bare – is used for wiring up two-way switching arrangements (see pages 84–5). Here the cores are coloured red, yellow and blue for identification only; in use, any or all may be live. It is known as **three-core-and-earth** cable.

Cable comes in a range of sizes, which are identified by the cross-sectional area of their conductors measured in square millimetres

Left to right: Two-core-and-earth cable is used for most household wiring work; the most common sizes are 6mm², 2.5mm² and 1mm²; three-core-and-earth cable is used for two-way switching arrangements. Single-core earth cable is used for earthing and cross-bonding

(mm²). The commonest sizes, and where you are likely to use them, are:

- 1mm² for lighting circuits
- 2.5mm² for ring circuits, 20-amp radial circuits and circuits to immersion and storage heaters
- 4mm² for 30-amp radial circuits (uncommon in domestic installations)
- 6mm² for circuits to showers and cookers.

Note that 30-amp radial circuits must be protected by a cartridge fuse or miniature circuit breaker (MCB), not by a rewirable fuse.

FLEX

Flex, like cable, contains cores within an outer sheath. The most widely-used type has three cores arranged in a triangular formation within a white or coloured round PVC sheath. All three cores are insulated: the **brown** core is **live**, the **blue** core is **neutral** and the **green-and-yellow** core is **earth**. The earth core allows an appliance or pendant light to be connected to the house earth continuity conductor (ECC) for safety.

You can also buy three-core flex with a rubber sheath that is covered with textile braiding (known as unkinkable flex), which should be used on domestic appliances such as irons, and heat-resisting flex which is used mainly to wire up immersion heaters and some types of enclosed light fittings.

Two-core flex has no earth core; it is used to wire up non-metallic light fittings and also appliances such as power tools and hair driers that are double-insulated and need no earth connection. The cores lie side by side, and are insulated and colour-coded as for three-core flex and contained in a round or flat PVC sheath (the latter is similar in shape to cable and is known as flat twin flex). Two-core flex insulated in clear or white PVC and left unsheathed (parallel twin flex, with a figure-of-eight cross-section) is obsolete.

Flex, like cable, comes in a range of sizes. The commonest are:

- 0.5mm² for small light fittings
- 0.75mm² for light fittings and small appliances rated at up to 1.3kW
- 1mm² for appliances up to 2.3kW
- 1.25mm² for appliances up to 2.9kW
- 1.5mm² for appliances up to 3.6kW
- 2.5mm² for appliances up to 5.7kW
- 4mm² for appliances up to 7.3kW.

Note that where flex is used to support a pendant light fitting, the maximum weight it can carry is 2kg (4½lb) for 0.5mm² flex, 3kg (6½lb) for 0.75mm² flex and 5kg (11lb) for all larger sizes.

Left, top to bottom: 1.5mm² three-core PVC-sheathed flex, unkinkable flex, two-core orange flex for use with double-insulated equipment outdoors, two-core round flex for pendant lampholders, two-core flat flex; right, top to bottom: curly flex, 1mm² three-core flex, bell wire, four-core telephone wire, coaxial cable for aerials

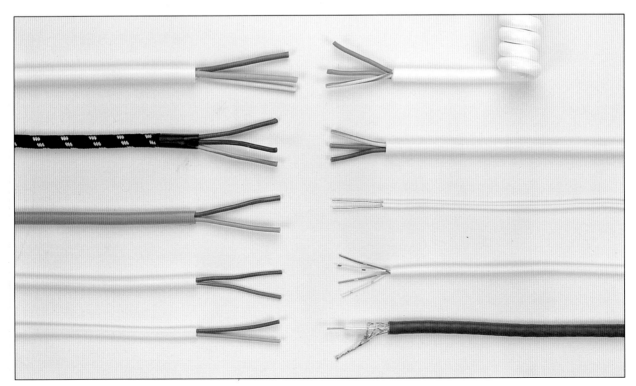

CONSUMER UNITS

The consumer unit is the heart of a modern domestic wiring system. It is a one-piece metal or plastic enclosure containing the system's main on-off isolating control. The incoming electricity supply reaches it via cables running from the meter, and is then distributed to the various lighting and power circuits in the house, via individual take-off points positioned along a metal strip within the unit called the **live busbar**.

Each circuit is protected against overloading – the drawing of more current than it is designed to supply – and against short circuits by the positioning of a protective device on the busbar at the take-off point of each circuit cable. Old-style units use **rewirable fuse carriers** or **cartridge fuses** as the protective devices, but now small automatic switches called **miniature circuit breakers** (MCBs) are used instead. The live core of the circuit cable is connected to the fuse or MCB; the neutral and earth cores go to separate terminals within the consumer unit.

The system may have a simple **on-off switch** as its main isolating control, but increasingly a **residual current device** (RCD) is being used. This not only serves as a manually operated main isolator switch; it also acts as an automatic safety device which cuts off the current flow in a fraction of a second if an earth fault develops.

RCDs are available with different current ratings and different sensitivities to current leakage. An RCD used as a whole-house protective device will generally have a current rating of 80 or 100 amps and a sensitivity to earth-fault currents of 100mA (milliamps). Such a unit will switch off quickly enough to provide protection against what is known as *indirect contact* – a shock received by touching something that should not be live but is, such as the metal casing of a faulty appliance. An RCD with greater sensitivity (normally 30mA) is required to provide protection against *direct contact* with live conductors or terminals. It may be subject to

Modern consumer units are available in a range of sizes. A split-load unit (left) is the norm for new installations, with only vulnerable circuits having RCD protection. Small units containing just one or two fuseways are ideal for extending existing systems

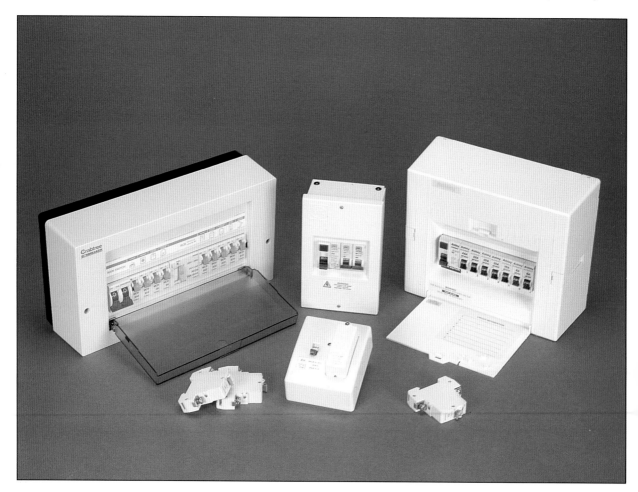

nuisance tripping if fitted as a whole-house RCD.

If the RCD is positioned on the consumer unit busbar to protect all the circuits in the house, all electricity is cut off when a fault develops on one circuit, unnecessarily leaving you without lights or power to the freezer, for example, and it contravenes the most recent Wiring Regulations. To prevent this problem an RCD may be used in conjunction with a main isolating switch so that the RCD protects only some of the circuits – in a **split-load consumer unit**. An alternative approach is to use a main isolator switch and to protect individual circuits with a device known as an **RCBO**. This is a manageable abbreviation for its full title: residual current breaker with overcurrent protection. An RCBO is essentially a combined RCD and MCB which provides overload, short-circuit and earth-fault protection in one unit. RCBOs are available with the same current ratings as MCBs (usually the Renard current values only), with sensitivities of 10mA and 30mA. Only the latter is generally used on domestic installations.

The drawback with both of these approaches is that other circuits then have no RCD protection; to provide it two RCDs can be used, one with 100mA sensitivity to protect the whole installation, and one with 30mA sensitivity to protect just the most vulnerable circuits. If the 100mA RCD is also used as the system's isolating switch, one incorporating a time-delay mechanism is necessary to prevent it operating when the more sensitive split-load RCD trips.

Consumer units and protective devices are designed on a modular basis, so that different combinations of equipment can be put together to suit individual systems. MCBs occupy one module, RCBOs, main switches and some RCDs occupy two units, and larger RCDs four. After selecting the various protective devices needed, it is then a simple matter to select a consumer unit with the appropriate number of modules to contain them all, ideally with room for one or two more circuits to be added if required in the future. Consumer units which can contain rewirable fuses, cartridge fuses, MCBs or a mixture of all three are also available. The smaller units, containing from one to six modules, are ideal for providing additional fuseways when extending an existing wiring system. Other components available for use in modern consumer units include timers, time-delay switches and bell transformers.

Current Ratings

MCBs and RCBOs are now commonly sized in accordance with the international Renard current ratings rather than the old Imperial ones, as part of a general move to European standardisation. The equivalents are as follows:

Imperial	Renard
5 amps	6 amps
15 amps	16 amps
20 amps	20 amps
30 amps	32 amps
45 amps	40 amps

Wiring Regulations

The use of RCDs in domestic wiring systems is generally discretionary; the user can choose to have enhanced levels of protection... at a price. However, there are two situations where the use of an RCD is essential to meet the requirements of the Wiring Regulations. The first is to protect any socket outlets which may reasonably be expected to supply equipment to be used outside the house. The second is to protect any circuits where the earth-fault current is not sufficient to blow the circuit fuse or trip the MCB within the appropriate disconnection time specified by the Regulations. Ascertaining whether this is so is a job for a professional electrician.

LIGHTING CIRCUIT FITTINGS

The main components on lighting circuits are the lighting points – ceiling roses or other outlets – and the switches that control them.

CEILING ROSES

A ceiling rose provides the connection point between the lighting circuit cable and a pendant light – either a lampholder carrying a lampshade or a suspended decorative light fitting. It consists of a circular plastic baseplate which is fixed to the ceiling surface, and a screw-on cover through which the pendant flex passes.

The baseplate has areas of thin plastic, known as *knockouts*, which are removed as required to allow the cable to enter from above the ceiling, and carries a bank of terminals to which the incoming cable and outgoing flex are connected. Most roses are designed for use with loop-in wiring (see page 40), and have separate terminals to accept the main circuit and switch cables. Others are for junction-box wiring (page 39), where just one cable enters the rose. All of them also contain an earth terminal.

LAMPHOLDERS

Lampholders hold lamps, which is the electrical trade's name for light bulbs. In a pendant light the flex from the rose passes through the lampholder cover and is connected to the terminals. Plastic lampholders have only live and neutral terminals; metallic ones also have an earth terminal and must be wired up with three-core flex. The metal socket into which the lamp fits is protected by a screw-on plastic skirt which also retains the lampshade if one is fitted; a special deep skirt (called a Home Office shield) must be used in bathrooms.

Batten lampholders are essentially a lampholder and ceiling rose in one, and are often used in bathrooms or as utility light fittings in lofts, garages and storerooms. The lampholder is mounted direct to its baseplate, which in turn is mounted on the ceiling or wall surface. Angled types are also available for wall-mounting.

LUMINAIRE SUPPORT COUPLERS

The drawback with conventional ceiling roses is that to remove the pendant light, the rose must be isolated from the mains supply and opened up to allow the flex to be disconnected. Luminaire support couplers (LSCs) get round the problem by providing a ceiling-mounted socket into which a special plug engages. It is then a simple matter to remove or replace the light for cleaning or during redecoration, without the need to turn the power off. The socket can replace a ceiling rose, or can be supplied as an adaptor to fit over the existing rose. It can also be flush-mounted if preferred.

A similar socket is also available for mounting wall lights, an increasing number of which are now available complete with the LSC plug already fitted to them. The socket can be fitted in a round conduit box or in a special shallow wall box.

There are several non–interchangeable plug–in coupler systems available at present. There are new British Standards governing the interchangeability of LSCs, so it seems likely that the earlier models will be phased out. Look for LSCs made to BS6972 and BS7001.

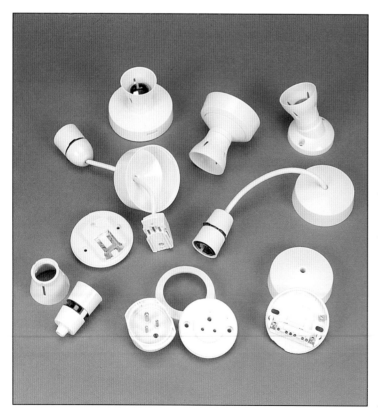

Top row: battenholders; bottom left: pendant lampholders; centre: LSCs; right: traditional ceiling roses

LIGHT SWITCHES·

The most widely-used type of light switch is the **plateswitch**. This contains one, two, three, four or even six rocker switches, known as *gangs*. Those with one, two and three rockers fit a square single mounting box, while those with four or six need a rectangular double box. You can also buy slim one-gang and two-gang **architrave switches**, designed for use on door architraves where space is limited. The switch cable is run from the lighting point to the switch controlling it and is then connected to the switch terminals on the rear of the face-plate. All can be mounted on shallow (16mm deep) mounting boxes, and will control fluorescent and low-voltage lighting arrays as well as tungsten-filament lamps.

A light controlled from just one position needs a **one-way switch**, which has two terminals on the back. If it is to be controlled from more than one position, **two-way switches** each with three terminals are fitted at each control position. The switch cable runs to one switch as before, and this is then linked to the other with three-core-and-earth cable (see pages 84–5). Two-way switches can also be used for one-way switching, with cable connected to just two of the three terminals. If control is needed from more than two positions, **intermediate switches** with four terminals are used between the two-way switches.

Plateswitches are available in several different finishes, from plain white or coloured plastic or metal to ornate brass in a range of period styles. They can be flush- or surface-mounted.

DIMMER SWITCHES

Dimmer switches are plateswitches that allow you to control the brightness of a light at the turn of a knob. On cheaper types you turn the knob back to 'zero' to switch the light off; more expensive types have a push-on/push-off action which avoids having to reset the light level each time the switch is turned on. One-gang and two-gang types are both available in a range of styles. They generally require a single square mounting box somewhat deeper than those used for plateswitches. Many have minimum and maximum wattage ranges, so it is important to match the switch carefully to the lighting load. Standard dimmers will not control fluorescent or extra-low-voltage lighting.

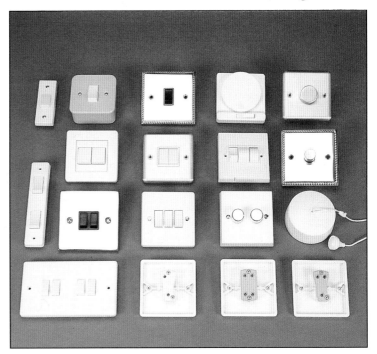

CEILING SWITCHES

In bathrooms a ceiling-mounted cord-operated light switch is mandatory if the switch is within reach of a bath or shower. Such a switch can also be used wherever the convenience of a cord-pull switch is wanted – to control bedside or garage lights, for example. Ceiling switches can be flush-mounted over a round conduit box, or surface-mounted on an integral mounting block; the latter may incorporate a neon light which can be wired up as an indicator to help locate the light switch in the dark. Separate types are available for one-way and two-way switching. White plastic is the standard finish.

SHAVER LIGHTS AND SOCKETS

Shaver lights – fluorescent or tungsten-filament striplights with a built-in shaver supply unit for an electric shaver – are often wired directly into lighting circuits. They have an integral pull cord which controls the light only, and can be wired up to a light switch if required. They are available in a range of sizes.

Shaver supply units and socket outlets may also be wired as a spur from a lighting circuit, or can be supplied via fused spurs from a socket circuit. See page 33 for more details.

Switches are available in a huge range of styles and finishes, including slim architrave switches (left) and dimmer switches (right). The three switch backs are (left to right) for one-way, intermediate and two-way switching

LAMPS AND TUBES

There is a huge range of lamps (the electrical trade's name for light bulbs) and tubes available for use in domestic light fittings. Those most widely used in the home can be divided into five main groups: general-service lamps, reflector lamps, compact fluorescent lamps, tubes and extra-low-voltage lamps.

Lamps have one of two types of end cap: the **bayonet cap** or the **Edison screw**. The former is pushed into a spring-loaded lampholder and twisted to engage the locking pins in the lampholder. It comes in two sizes: 22mm (known as BC for short) and 15mm (small bayonet cap or SBC). Edison screw (ES) caps are screwed into special threaded lampholders and come in a larger range of sizes, but most fittings either take the standard ES or the small SES size.

GENERAL LIGHTING SERVICE (GLS) LAMPS

This is the most widely-used type of lamp in most homes. The glass envelope (the 'bulb') may be clear, pearl (translucent), white or coloured. White lamps with just a hint of colour have become increasingly popular in recent years. The standard pear and mushroom shapes have now been joined by lamps with a slightly squared-off profile.

GLS lamps also include more exotic shapes such as pointed candle lamps with smooth or twisted clear, white or coloured glass envelopes, compact 'pygmy' lamps and a growing range of so-called 'decor' lamps with round white or coloured envelopes designed to be on

show when used with ornamental lampshades or fittings.

GLS lamp wattages range from 25W to 150W, with 40W, 60W and 100W sizes the most common in the home. Pygmy lamps are usually rated at 15W, candle lamps at 25W or 40W only.

REFLECTOR LAMPS

Reflector lamps are designed for use in spotlights and floodlights, where a high-intensity directional beam is required rather than a diffuse source of light. Part of the envelope is coated with silver on the inside to reflect the light in one of two ways.

The **crown-silvered** (CS) reflector is usually a pear-shaped lamp with the top of the envelope silvered so that light is thrown back into a curved reflector; this then projects the light forward as a narrow beam – commonly 20–25° wide.

Internally-silvered lamps are silvered round the base and sides of the envelope and throw the light forward without the need for a curved reflector. Those with a standard flattened pear profile produce a wider beam than CS lamps, but specially shaped versions with a narrower crown can rival the CS beam angle.

PAR (parabolic aluminised reflector) lamps have pressed rather than blown glass envelopes, and are shaped like early space capsules. They are much tougher than blown glass lamps, and can be used out of doors. Spot types have a very narrow beam – usually 10-12° – while flood types produce a beam about 30° wide. PAR lamps are also available with coloured glass lenses.

Reflector lamps are available in the same range of wattages as GLS lamps. An electronic PAR halogen lamp with a lower wattage rating and current consumption than standard types is now also available.

COMPACT FLUORESCENT LAMPS

These are variations on a technologically cunning theme of combining a miniature fluorescent tube and its control gear in a lamp small enough to fit a standard lampholder in a typical lampshade or fitting. There are several types, including lamps with a cylindrical or spherical glass envelope and the cluster lamp, which has four small glass fingers projecting from the lamp base. All have the twin advantages of using around a fifth as much electricity as a standard GLS lamp (wattages range from 9W to about 25W), and promising a lamp life of about

GLS lamps come in a wide range of shapes and sizes. Top row: standard pear shape; middle row: mushroom, squared-off pear and globe shapes; bottom row: candle and pygmy shapes

8,000 hours – eight times as long as the GLS average. They do, of course, cost more to buy, but over time they cost less than half as much to run and replace.

TUBES

Fluorescent and filament lamps in tubular form (also called strip lights) come in a wide range of sizes and wattages. **Standard fluorescent tubes** are 38mm in diameter and come in five standard sizes (600mm/20W, 1200mm/40W, 1500mm/65W, 1800mm/85W and 2400mm/125W). **Slimline tubes** 28mm in diameter come in the same sizes but with wattages about 10 per cent lower. **Miniature fluorescent tubes** range in size from 150mm/4W up to 525mm/13W. All have standard bi-pin end caps. Lastly, **circular fluorescent lamps** are also available in sizes up to 410mm diameter and in wattages from 22W up 60W. Fluorescent lamps offer a range of shades of 'white' light, from a cold bluish hue to a warm pinkish effect.

Tungsten-filament strip lights are smaller – generally between 220mm and 285mm long – and are mainly used for concealed lighting or for decorative lighting over mirrors, in display cabinets and above aquaria. The tube may be clear or pearl, and wattages range between 30W and 60W.

EXTRA-LOW-VOLTAGE LIGHTING

Extra-low-voltage lighting uses miniature 12-volt halogen lamps run from a transformer to provide localised lighting via a range of neat recessed downlighters, eyeball fittings or wall washers and surface-mounted spotlight fittings. The lamps themselves are shaped like small PAR reflectors, and give a cool, white light that is ideal for highlighting specific features in any room. Common wattages are 20W and 50W and lamp life is around 2,000 hours. For safety's sake extra-low-voltage lighting must be run from a special transformer made to BS3535, designed for the job and positioned where it cannot overheat. The safest type to use for d-i-y installations is a multiple spot or track fitting complete with its own built-in transformer, which is designed for connection to an existing ceiling rose or junction box and which will be controlled by the existing light switch.

Above: Reflector lamps

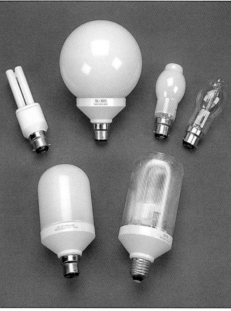

Left: Compact fluorescent lamps. Top right are low-wattage mains-voltage halogen lamps, which provide halogen-quality light without the expense of the new luminaires and transformers in other systems

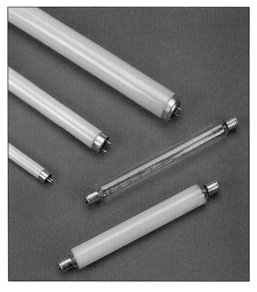

Left: Fluorescent and filament tubes

SOCKET CIRCUIT FITTINGS

Socket circuits supply power to the wide range of small appliances found in every home. These appliances are either linked to the mains via a fused plug pushed into a socket outlet, or else are wired in permanently via a fused connection unit.

PLUGS

A plug connects the flex from an appliance to the mains. Inside the plug, the flex cores are screwed to their respective terminals. On a modern wiring system using plugs with rectangular pins, the neutral and earth terminals are part of the pins themselves, while the live terminal is physically separated from the live pin by a small cartridge fuse. This is rated at 13 amps for appliances consuming up to about 3 kilowatts, and at 3 amps for those taking less than 720 watts. This type is commonly referred to, nevertheless, as a 13-amp plug.

On older wiring systems the plug is unfused and has round pins; the fusing is provided at the start of the circuit supplying the socket. The plugs come in three sizes, rated at 2, 5 and 15 amps.

All 13-amp plugs are broadly the same in principle, but differ in detail. The terminals may be a threaded pillar with a screw-down stud, a hole with a small terminal screw or a snap-down terminal clip. The flex grip may be a screw-down bar or some form of nylon jaw or vice. The pins and the plug-top screw may or may not be captive. The inside of the plug may

be designed so it can be wired with all the flex cores cut to the same length. Lastly, on new plugs the live and neutral pins must now for safety reasons be shrouded for part of their length with insulating material, which may be colour-coded – brown for the live pin, blue for the neutral.

Many appliances are now sold with moulded-on 13-amp plugs. These allow the fuse to be changed if necessary via a retractable fuse carrier, but cannot be opened. If they need replacing they must be cut off so the flex can be stripped and reconnected to a new plug. The old plug should have its pins bent before it is thrown away so it cannot be inserted into a socket outlet.

SOCKET OUTLETS

Modern wiring systems use socket outlets with rectangular slots into which a 13-amp plug fits. They are by far the commonest wiring accessory in the average house, which may have 30 or 40 outlets.

Modern socket outlets are available in single, double and triple versions, able to accept one, two or three 13-amp plugs respectively. Each outlet may be switched or unswitched, and may also have a small neon indicator to show whether it is on or not. They can be fitted on surface-mounted plastic or metal boxes (the latter mainly in utility areas such as garages or workshops). Alternatively, they can be flush-mounted in plastic or galvanized steel boxes recessed into the wall surface. Old-style sockets taking round-pin plugs are available only as single sockets; the 5-amp and 15-amp versions may be switched or unswitched, but the 2-amp type is usually unswitched.

Socket outlets can also be mounted in suspended timber floors to avoid having trailing flexes running from wall socket outlets to an appliance on furniture in the centre of the room. Floor socket outlets have a spring-loaded cover flap that keeps dust and debris from getting into the slots.

FUSED CONNECTION UNITS

A fused connection unit (FCU) provides a connection between the mains and an appliance that needs to be permanently connected, such as a freezer or a wall-mounted fire. It can also be used to provide the required overload and short-circuit protection on a sub-circuit taken off the main socket circuit – to supply a wall light, an electric towel rail or an extractor fan, for example.

Plugs, adaptors and flex connectors. Left, from top to bottom: three-way adaptor, shaver adaptor, fixed flex connectors, two-part flex connector; top row: four-way wire-in adaptor, time switch; centre, clockwise from top: easy-pull plug, round-pin plugs, standard 13-amp plugs

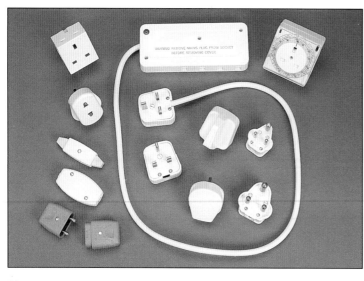

Left: Socket outlets come in a number of styles, switched or unswitched, with or without a neon on-off indicator light. Bottom left: 2-amp round-pin sockets are useful for circuits to table lamps. Bottom right: RCD-protected outlet for powering appliances used outdoors

Below: Fused connection units may be switched or unswitched with or without neon indicators, and may have a flex or cable outlet

In the former situation, the flex from the appliance is connected directly into the FCU through an aperture in its faceplate. A flex grip holds the flex sheathing and guards against the flex being pulled and breaking its cores inside the unit. A small fuse carrier holds the appropriate cartridge fuse, providing the same protection against overloading as a plug fuse does at a socket outlet.

When an FCU is used on a sub-circuit, cable is usually run on from it to the appliance concerned. It may be connected directly to the appliance, or may terminate near the appliance at another accessory called a **flex outlet plate**; this contains two sets of terminals, allowing a length of flex to be used for the final connection between the sub-circuit cable and the appliance.

FCUs are the same size as a single socket outlet, and can also be mounted side by side in pairs on special dual mounting boxes. They may be switched or unswitched, and like socket outlets may have a neon indicator light. Switched FCUs should have double-pole switching, so that both the live and neutral supplies are cut when the switch is off; this means that the appliance can be completely isolated from the mains for repairs without the need to disconnect the appliance flex.

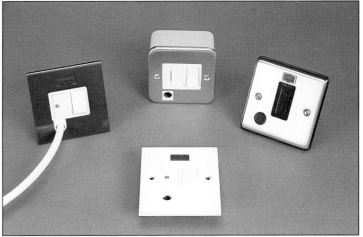

DOUBLE-POLE SWITCHES

Another situation where double-pole switching may be used in conjunction with socket outlets is in fitted kitchens, where unswitched socket outlets are now often fitted at the back of recesses beneath worktops for slide-in appliances such as fridges or dishwashers. Since it is clearly impractical to have to pull the appliance out to turn off its main power supply, the supply to the socket is controlled by a separate double-pole isolating switch fitted above the worktop. See pages 32-3 for more details.

OTHER POWER CIRCUIT FITTINGS

Apart from supplying power to appliances via socket outlets and FCUs, power circuits also feed larger current users such as electric cookers, water heaters and showers direct, generally via a separate circuit leading to each appliance. The appliance itself must be capable of being completely isolated from the mains supply for repair or servicing, so the supply to it is interrupted by a switch that breaks both the live and neutral sides of the circuit - a **double-pole** (DP) switch.

Above: Cooker control units, cooker switches and a cooker connection unit

Right: Other double-pole switches

WALL-MOUNTED SWITCHES

Wall-mounted DP switches are available in a range of current ratings, from 20 up to 50 or even 63 amps. The 20-, 32- and some 40- and 45-amp switches are the same size as a single socket outlet; others are the size of a double socket. All are designed to accept an incoming circuit cable and an outgoing cable to the appliance itself, and can be flush- or surface-mounted. In addition, the 20-amp size is also available with a flex outlet for use with appliances such as immersion heaters and wall-mounted electric fires. All are available with a neon indicator to show that the power supply to the appliance is on. Most manufacturers offer faceplates printed with the words 'WATER HEATER' or 'COOKER' as appropriate, and one range offers a choice of no less than 24 different markings, which is very useful in a modern kitchen with many permanently wired-in appliances. It is also possible to buy a special dual DP switch to control twin-element immersion heaters.

Cooker control switches are also available as so-called cooker control units, which include a 13-amp single socket outlet. These have a 45-amp double-pole switch controlling the cooker and a separate switch for the socket outlet. Apart from supplying a socket outlet that costs nothing extra to install, it is difficult to see the advantage that this unit provides; it has the definite drawback that if the unit is sited close to

the cooker or hob it is controlling, the flex leading to the socket outlet from a portable appliance could trail across a hot burner or hotplate and be damaged by the heat. It is therefore generally safer to fit a separate DP switch for the cooker, and to site socket outlets well away from the hob position. The unit, as with other DP switches, can be flush- or surface-mounted.

CEILING-MOUNTED SWITCHES

DP switches are also available in ceiling-mounted designs, intended mainly for use in bathrooms, where wall-mounted switches must be out of reach of anyone using the bath or shower – a requirement hard to meet in many small modern bathrooms. They are available in two current ratings – 15- or 16-amp for controlling appliances such as wall heaters, and 40- or 45-amp for controlling electric showers. Both usually have a neon light to show when the appliance is on, and the latter type must also have a mechanical on-off indicator as back-up in case the neon indicator lamp should fail. Both flush- and surface-mounted versions are available.

SHAVER SUPPLIES

The mains electric shaver is the only appliance for use in the UK that has a two-pin plug. It can be used with a special shaver adaptor that fits a 13-amp socket outlet, but this is a poor solution for two reasons. The first is that 13-amp outlets are not allowed in bathrooms, where shavers are generally used. The second is that a typical shaver flex is fairly short, and may not reach a socket outlet easily, especially if it is at skirting-board level. The solution is to fit a special outlet just for shavers at a convenient position for the user.

In a bathroom or washroom, an outlet called a **shaver supply unit** must be used for safety reasons, because of the proximity of water. This unit (which must be made to BS3535) contains a transformer and provides an earth-free power supply to the shaver that is completely isolated from the mains. Most units have two sets of sockets, one supplying an output of 110/115 volts and one of 230/240 volts, so shavers from countries with lower mains voltages than the UK can also be plugged in. The unit is protected by a self-resetting thermal overload device, which restricts the power that can be drawn from it and so prevents other appliances being used.

In other rooms, a smaller **shaver socket outlet** can be fitted instead, perhaps by the bed or close to a dressing table. This does not contain a transformer, so is much cheaper than a shaver supply unit, but is protected against misuse by a self-resetting overload device and also by a 1-amp fuse.

Both types can be flush- or surface-mounted, and can be supplied from either a power or a lighting circuit; see page 101 for more details.

Top: the shaver supply unit contains an isolating transformer and is the only socket outlet allowed in a bathroom or washroom. Right: the shaver socket outlet can be fitted anywhere else. Left: A plug-in shaver adaptor is a useful standby

NON-MAINS FITTINGS

This is a group of fittings for indoor use manufactured to match the style of other accessories
such as switches and socket outlets, but which
are not part of the house wiring system. They
provide connection points for telephones, TV
and radio aerials and computer equipment. All
are the same size as a single socket outlet, and
can be flush- or surface-mounted.

TV AND FM COAXIAL SOCKET OUTLETS

The coaxial downlead from a TV or FM aerial
terminates at a special socket outlet, into which
the lead to the receiver itself is then plugged. A
single outlet can be fitted wherever required in
the house, or separate TV and FM radio downleads can be run side by side to a double outlet, allowing both receivers to be plugged in at
one location.

An alternative arrangement that will supply
both TV and FM connection points is to use a
pair of special isolated double socket outlets,
each containing a device called a **diplexer**.
One outlet is fitted in the loft, close to the aerials, and the other wherever it is required inside
the house. The aerial leads are plugged into
their labelled outlets in the loft, and a single
coaxial downlead is then run to the house outlet. There the two receivers are plugged in to
the appropriate outlets.

This arrangement is neater than one with
separate downleads for each aerial, but will
cost rather more unless the coaxial cable run is
very long.

TELEPHONE SOCKET OUTLETS

As with aerial socket outlets, it is obviously
desirable to have telephone outlets that match
other domestic wiring accessories. Following
the liberalisation of the rules concerning fitting
telephone socket outlets in the home, it is now
possible for the householder to add extra outlets so long as the **master outlet** and the
incoming line wiring to it is carried out by a
British Telecom engineer or a BSI (British
Standards Institution) approved installer. A special adaptor is used to connect the **secondary
outlet**(s) to the master outlet, and special telephone cable similar to round flex is then run
on to the extra sockets.

Master and secondary telephone outlets are
available in one- and two-gang versions. Make
sure that the correct type is chosen for use as a
secondary outlet.

COMPUTER TERMINAL SOCKETS

Most wiring accessory manufacturers now offer
a range of computer terminal socket outlets to
match their mains-voltage fittings. These are
available in one-and two-gang types, in different versions to accept the various plugs in common use, including BNC, RJ45, 601W and 25-
pin D-type connectors. In addition, some
ranges include a combined computer terminal
socket and a single master telephone outlet in
one standard-sized unit.

*Fittings for TV and
radio aerials and tele-
phones are available to
match other wiring
accessories*

FITTINGS FOR OUTDOOR USE

Electricity can be a very useful servant out of doors as well as within the house, supplying light and power to both the immediate surroundings such as the patio and further afield to outbuildings and garden ponds. Its biggest enemy is obviously wet weather, which means that the installation and any fittings used on it must be weatherproof.

British Standard EN60529 details a rating system for classifying the resistance of electrical fittings to dust and water penetration. These IP (index of protection) ratings consist of two digits; the first runs from 0 to 6 and refers to protection of persons from contact with live or moving parts and resistance to the ingress of solid objects such as dust particles, while the second runs from 0 to 8 and covers resistance to the ingress of water. The minimum figure suitable for outdoor use is an IP54 or IP55 rating, which means that the fitting is protected against water splashed from any direction. An IP65 rating means protection against low-pressure water jets (as well as total dust protection).

OUTDOOR SWITCHES

There are several types of switch suitable for use out of doors. One has a plastic casing containing a rocker switch sealed behind a flexible plastic membrane; another has a metal case and a rotating on-off switch. Both are surface-mounted, rated at 6 amps, available with one- or two-gangs in versions suitable for one-way or two-way switching, and are designed to be used in conjunction with round conduit to protect the cable run to the switch.

In addition to conventional on-off switches, **photoelectric switches** can be fitted to turn outside lighting on at dusk and off at dawn (or, with adjustable types, at other pre-set light levels).

Passive infra-red (PIR) detectors sense the presence of heat emitters such as visitors or intruders within their detection range, and switch lighting under their control on immediately and off again after a pre-set time. They save energy and offer both convenience and security.

OUTDOOR SOCKET OUTLETS

Socket outlets outside the house, whether on the house wall or in the garden, make it much easier to use power tools such as lawnmowers

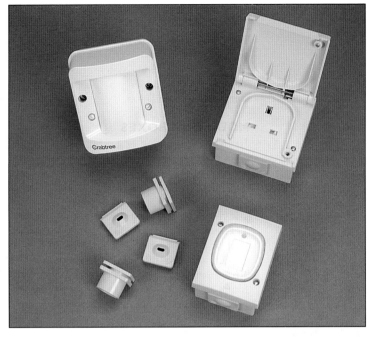

Weatherproof fittings for outdoor use. Clockwise from top left: PIR detector, socket outlet, outdoor switch

and hedgetrimmers, since there is no need to trail a long extension lead out through a window or door from an outlet inside the house. Outdoor socket outlets usually have a hinged spring-loaded flap which covers and seals the pin holes against water when the socket is not in use. Most will accept an ordinary 13-amp plug, but some require a special plug that interlocks with the socket outlet to provide a weatherproof connection. As with outdoor switches, they are designed for use with round conduit. One- and two-gang versions are available, usually only in unswitched form.

Socket outlets outside the house must be protected by a residual current device (RCD). This can either be wired into the circuit supplying the socket(s), or can be incorporated within the socket outlet itself. The RCD used must be a 10 or 30mA type.

OTHER OUTDOOR FITTINGS

Any light fittings installed outside must obviously be suitable for the purpose. Always check that fittings are clearly labelled for outdoor use.

Extra-low-voltage equipment such as garden lighting and pond fittings is becoming increasingly popular, both for safety reasons and because of comparatively lower running costs when compared with mains-powered fittings. The transformer must be made to BS3535 and must normally be sited indoors or in a weatherproof chamber outside.

ANCILLARY ITEMS

The various groups of wiring accessories described so far in this chapter are the visible parts of a house wiring system. However, there are several other hidden components needed to complete the installation.

MOUNTING BOXES

To satisfy the Wiring Regulations, the connections between circuit cables and wiring accessories have to be made within a non-combustible enclosure. This is provided by a mounting box.

If the accessory is to be flush-mounted, a recessed mounting box made from galvanised steel or plastic is used. The box must be deep enough to accept the accessory, and galvanised boxes come in different depths ranging from 16mm to 47mm. Plastic boxes are generally available in only one depth - 35mm - although some ranges include shallower boxes designed for use with light switches.

For surface-mounting, boxes are either of white plastic or moulded metal, the latter usually having a paint finish and intended for use with utility-type metal-clad wiring accessories. Standard box depths range from 19mm to 46mm.

Boxes are available in one-gang and two-gang sizes, with some accessory ranges also offering special boxes to accommodate a triple socket outlet and to accept two one-gang accessories side by side. Most ranges also include larger or deeper flush and surface boxes, specially designed to accept cooker control units and shaver supply units respectively.

Some square accessories can also be ceiling-mounted using flush or surface wall boxes. Round-based accessories can either be surface-mounted over a round surface box, or flush-mounted over a round conduit box screwed to the underside of a batten fixed between the joists. Both these boxes have two built-in fixing lugs 51mm apart, allowing the accessories to be secured direct to the box.

Mounting boxes have pre-punched discs (metal) or weakened areas (plastic) known as *knockouts*, designed to be removed to allow the cable to enter the box. In metal boxes the resulting hole must be lined with a rubber grommet to stop the cable sheath from chafing. Metal boxes all have an earthing terminal fitted, and most also have one adjustable lug to allow the accessory faceplate to be set level even if the mounting box is not. They all have pre-formed fixing holes in the base of the box.

JUNCTION BOXES

Connections between circuit cables other than those in wiring accessories must also be made in a non-combustible enclosure, and fittings called junction boxes are used for this. They are generally hidden in floor and ceiling voids, and are made of brown or white plastic. The box is round, with a screw-on lid that can be rotated to reveal a number of cable entry points. There are three commonly used types, with three, four and six terminal blocks respectively. The first is used for connecting cable lengths together and for wiring up spurs off main circuits. The other two are used to connect in switch cables on lighting circuits. A rectangular white three-terminal junction box is available for wiring spurs taken from power circuits that are surface-mounted.

EARTH SLEEVING

The current-carrying cores of cables are insulated, but the earth continuity conductor is bare. When it is exposed within a wiring accessory, it must be covered with a length of green-and-yellow PVC sleeving.

Metal and plastic mounting boxes are available in different depths to suit individual accessories. The flanged plastic boxes at the bottom are used to mount accessories in partition walls

CABLE CLIPS

Cable that is surface-mounted or fixed to the sides of joists is secured with plastic cable clips. These come in a range of sizes to match the various cable types used, and have a hardened steel fixing pin that can be driven into masonry or wood.

MINI-TRUNKING, CONDUIT AND CHANNELLING

Surface-mounted cable can also be run in square or rectangular **mini-trunking**. The U-shaped base section is pinned or stuck to the wall or ceiling surface, and the lid is clipped on once the cable runs are in place. Mini-trunking comes in a range of sizes from 16mm square to 50x32mm, and there is a full range of couplers, angles and T-fittings available to connect the lengths together.

Concealed cable runs were simply buried in plaster, but on new work it is mandatory either to feed the cables through **oval PVC conduit** which is held against wall surfaces with galvanised nails, or to cover the cable runs with lengths of pinned-on **PVC** or **metal channelling**. Both are then plastered over.

Outdoor cable runs on walls or buried underground can be protected by running the cable(s) inside **round PVC conduit**. As with trunking, a range of fittings is available; these are *solvent-welded* in the same way as plastic plumbing fittings. Round conduit can also be used within floor and ceiling voids, but is seldom found on domestic wiring installations.

All three products are supplied in standard 2m or 3m lengths, and can easily be cut down with a hacksaw or other fine-toothed saw.

Above: cable clips, strip connectors, plug fuses, plastic grommets and green/yellow earth sleeving

Left (from top): Round conduit mainly used to protect cable runs out of doors, oval conduit and channelling used to protect cable runs buried in plaster, mini-trunking – for concealing surface-mounted cable runs – and fittings for connecting it to a mounting box, junction boxes for connecting cables

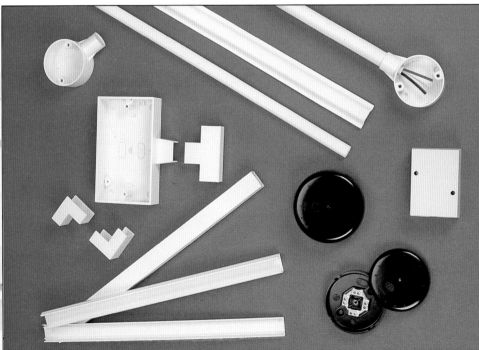

LIGHTING CIRCUITS

Lighting circuits are wired as **radial circuits**, with the cable running from the consumer unit to the first lighting point and then on to other points on the circuit in series, terminating at the most remote point. A house should have at least two lighting circuits, so that it is not plunged into darkness if one lighting circuit fuse or MCB is activated. These are normally wired as upstairs and downstairs circuits; in larger homes it is common to have two downstairs lighting circuits to cope with the likely demand. There is no limit to the floor area that can be served by each circuit (but see the paragraph below about cable sizes).

A lighting circuit is normally protected by a 5- or 6-amp circuit fuse or MCB, which means that it can supply a maximum load of 1200 or 1440 watts respectively. (Amps x volts = watts.) Each lighting point is assumed to have a power demand of 100 watts, so the circuit can in theory supply up to 14 separate lighting points. In practice, it is usual to restrict each circuit to about eight points, to allow for the use of lamps more powerful than 100 watts or for the installation of multi-lamp fittings such as chandeliers or spotlamp clusters at certain lighting points.

The circuit is usually wired using 1mm² two-core-and-earth PVC-sheathed cable. However, if the circuit is very long, it may be necessary to use 1.5mm² cable to avoid excessive drop in the mains voltage: this is caused by the resistance of the cable and becomes more pronounced the longer the cable run is. If two-way switching is involved, three-core-and-earth cable is used to connect the two-way switches together.

SWITCH CONTROL

Each lighting point is generally controlled by its own switch, although there are obviously situations where more than one light - a set of wall lights, for example - is controlled by a single switch. The switch is wired into the live side of the circuit so that only the switch feed - the red core of the switch cable - is permanently live.

The light switch may be wall- or ceiling-mounted. Ceiling switches are operated by a pull cord, and are used for convenience - to control bedside or garage lights, for example - or, in bathrooms, for safety. A wall-mounted switch must not be sited within reach of anyone using a bath or shower (reckoned to be a notional 2.5m), and in many of today's small bathrooms such a switch would be well within reach. An alternative solution to using a ceiling switch in this situation is to put a wall switch outside the bathroom.

INSTALLING LIGHT FITTINGS

A ceiling rose makes a safe connection between the circuit wiring and the flex running to a pendant lampholder. Light fittings other than pendant lampholders must be installed over a non-flammable box which contains the connections between the fitting and the circuit wiring. For ceiling lights a round conduit box recessed into the ceiling is ideal. For wall lights a conduit box is the first choice, although a smaller architrave box may be better suited to fittings with small baseplates. Luminaire support couplers (LSCs) are an efficient modern alternative, allowing the light fitting to be removed, for example when redecorating, without exposing any of the wiring.

With junction box wiring, separate cables to the light fitting and its switch are connected into the main circuit cables using a four-terminal junction box. The black core of the switch cable is tagged with red tape to indicate that it is live

JUNCTION-BOX WIRING

In the early days of household electricity, lighting circuits were wired up using porcelain and brass fittings. The circuit cable, which was not earthed, ran from the fusebox to a series of junction boxes, and separate lengths of cable connected each box to a ceiling rose and its associated switch. The circuit terminated at the final junction box.

Junction-box wiring is still used today, but with two main differences: all lighting circuits are now earthed throughout, and the accessories used are plastic, not porcelain.

To wire up a four-terminal box, the circuit and switch cable live cores are taken to terminal block 1. The switch cable neutral core and the live core of the cable feeding the ceiling rose or light fitting are taken to terminal block 2. Since the switch cable neutral core is live when the switch is on, it is essential to identify it as such by wrapping its insulation in red PVC insulating tape. The circuit and light cable neutral cores are taken to terminal block 3 and the earth cores of all the cables are taken to terminal block 4. All the bare earth cores are first covered with lengths of green-and-yellow PVC sleeving.

Junction-box wiring is seldom used as the sole means of wiring up a modern domestic lighting circuit nowadays, thanks to the widespread use of loop-in roses (see over). However, a junction box is often used to extend an existing lighting circuit, and in some instances junction-box wiring of part of the circuit may make for more economical use of cable.

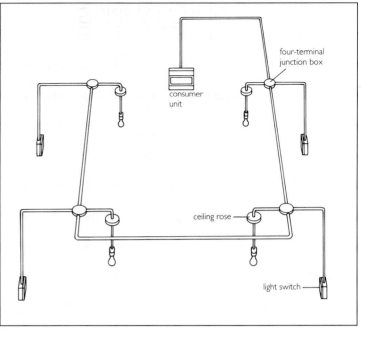

four-terminal
junction box

consumer
unit

ceiling rose

light switch

Above: The circuit cable runs from junction box to junction box, with separate cables running to each light and switch

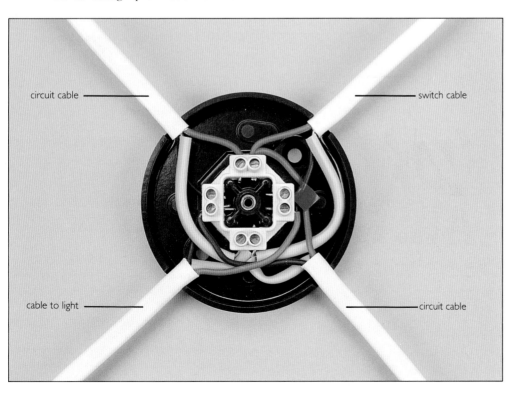

circuit cable

switch cable

cable to light

circuit cable

LOOP-IN WIRING

Loop-in wiring was developed from the use of junction boxes to connect the circuit cable and those supplying each lighting point and its switch in one enclosure. By adding a third terminal to the traditional two-terminal (two-plate) ceiling rose, it became possible to 'loop' the circuit cables directly in and out of successive roses along the circuit, making the junction box redundant. The three-plate ceiling rose, although still available, has been developed into the multi-way terminal block generally used today.

To wire up a loop-in rose, the circuit and switch cables are brought in through a central knockout in the rose's baseplate. Their live cores are taken to the central terminal block, which has a screw-down terminal for each of the three cores. The circuit cable neutral cores are taken to the end block with two screw-down terminals. The switch cable neutral core is taken to the single terminal at the other end of the strip, and is identified as being live with red PVC tape. All the earth cores are covered in green-and-yellow PVC sleeving before being connected to the separate earth terminal.

The live and neutral flex cores for the light are connected to the terminals at the ends of the main terminal strip: the live (brown) core next to the switch cable neutral core, the neutral (blue) core next to the terminal containing the circuit cable neutral cores. If three-core flex is being used, the earth core is connected to the earth terminal with the cable earth cores. The live and neutral flex cores are looped over restraining hooks before the rose cover is fitted.

These hooks take the weight of the pendant lampholder and any lampshade fitted. 0.5mm² flex can support a weight of 2kg (4½lb), 0.75mm² flex up to 3kg (6½lb) and larger sizes a maximum weight of 5kg (11lb).

Loop-in circuits can be extended by adding a spur cable connected in to the rose's live and neutral terminal banks; the spur cable earth core goes to the separate earth terminal. Alternatively, a junction box can be used by wiring it into the circuit at any convenient point.

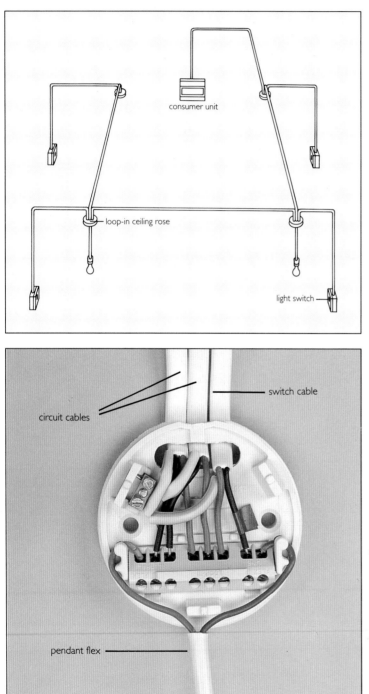

consumer unit

loop-in ceiling rose

light switch

switch cable

circuit cables

pendant flex

Above left: The circuit runs from loop-in rose to loop-in rose, with cable run from the rose to the switch and flex run to the lampholder

Left: With loop-in wiring, the main circuit cables loop in and out of each rose. The switch cable is connected to the live terminal and the switch return terminal, where the black cable core is tagged with red tape to indicate that it is live. Loop-in wiring can also be used to connect up other decorative light fittings which have hollow baseplates or are mounted over a recessed conduit box. Strip connectors are used to replicate the connections made within the rose

EXTRA-LOW-VOLTAGE LIGHTING

Extra-low-voltage lighting runs at 12 rather than at 240 volts. It is commonly referred to simply as low-voltage lighting, although that is actually the correct technical term for mains voltage. The light fittings are supplied from a transformer which is itself supplied at mains voltage. The transformer, which must be made to BS3535, may be remote from the fittings it supplies, perhaps concealed within the ceiling void, or may be an integral part of the fitting itself.

The special light fittings required for low-voltage lighting use small tungsten-halogen lamps which emit up to three times as much light as a conventional lamp; for example, a 50-watt halogen lamp is the equivalent of a 150-watt filament type. The light emitted is a cool white, ideal for highlighting room features as well as for creating other lighting effects, and minimises ultra-violet fading of sensitive materials. The lamp can be expected to last up to three times as long as a filament lamp, and overall energy consumption can be reduced by as much as 60 per cent.

The actual circuit wiring could not be simpler. The supply to the transformer is run in 1.5mm² two-core-and-earth cable, ideally taken from a separate 5/6-amp fuseway in the consumer unit. If this is not possible, a spur can be taken off an existing circuit; this must run via a fused connection unit fitted with a 5-amp fuse if it is taken off a power circuit. The cable is then taken to a four-terminal junction box, where the cable run to the switch controlling the lighting is connected. From here, the cable is taken to the transformer and then on to the terminal block of each fitting, but the earth core is not used from the transformer onwards since extra-low-voltage lighting is not earthed.

The size of cable used to connect the light fittings to the transformer depends on the lamp wattage and the total length of the cable run. Always check the lighting manufacturer's specifications for cable sizes when installing such a system.

If an existing ceiling fitting is being replaced by extra-low-voltage lighting, the rose can be re-used as a junction box for connecting the transformer wiring. The rose should be screwed to the side of a ceiling joist.

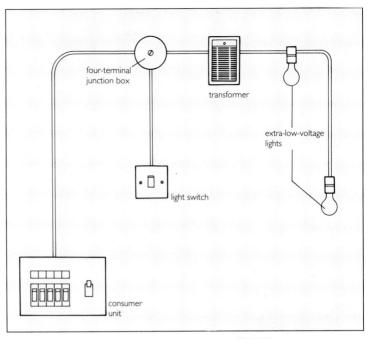

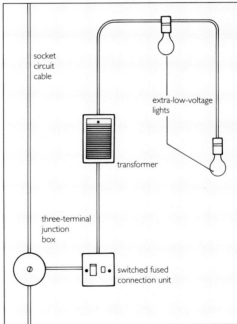

The power supply for an extra-low-voltage lighting circuit is best taken from a 5/6-amp fuseway in the consumer unit (shown above). The cable runs first to a junction box, where the light switch is wired in, and then on to the transformer supplying the lights. Left: An alternative is to take a spur off an existing circuit. If a power circuit is used, the spur must run via a fused connection unit; this can also act as the on-off switch for the sub-circuit

SOCKET OUTLET CIRCUITS

Power circuits supplying socket outlets and fused connection units (FCUs) can be wired in one of two ways; as **ring circuits** or as **radials** (like light circuits). Most homes will have at least two power circuits, either one for upstairs and one for downstairs or one for each half of the house, and it is common practice in new homes to provide a separate circuit to the kitchen where a large number of appliances is often in use. Fixed appliances or equipment with high current demands, such as space heaters, water heaters and immersion heaters, should not be supplied from socket circuits, but should have their own separate circuits (see pages 46 and 48).

The numbers of socket outlets fitted in the home is a matter of personal choice, but since modern power circuits can supply an unlimited number of outlets it makes sense to have too many rather than too few. Their distribution has been the subject of numerous recommendations since the Parker Morris Report of 1961 (which suggested a minimum of 15 single outlets, a total that would be woefully inadequate even in a small flat today). A desirable number and distribution of double outlets in a modern home is as follows:

Kitchen	6
Living-room	6
Dining-room	4
Bedroom	4
Hall	1
Landing	1
Utility room	2
Garage	2

The figures above do not include any FCUs used to supply appliances.

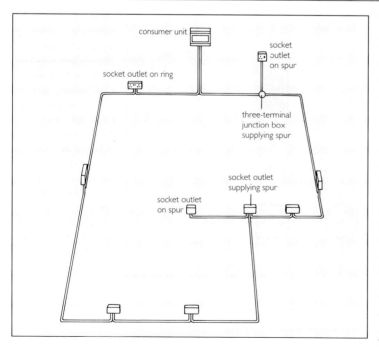

A ring circuit consists of a continuous loop of cable running out from and back to the consumer unit, supplying socket outlets either directly or by means of spur cables

RING CIRCUITS

The ring circuit, as its name implies, is wired up as a continuous ring. The circuit cable's live core starts and terminates at the fuse or MCB protecting the circuit, allowing current to flow round the circuit in either direction and effectively increasing the circuit's current-carrying capacity without the need for larger and more expensive cables. The neutral and earth cores are connected to their respective terminal blocks in the consumer unit or fusebox.

Since it is unlikely that all the socket outlets on the circuit will be in use at the same time, no limit is placed on the number of outlets the circuit can supply. The only restrictions imposed by the Wiring Regulations are that each circuit should serve rooms with a floor area of not more than 100sq m (1075sq ft), and that the number of unfused spurs taken off the circuit (see page 44) should not exceed the number of sockets or FCUs on the main circuit. The circuit is wired in 2.5mm^2 PVC-sheathed two-core-and-earth cable, and is protected by a 30- or 32-amp fuse, MCB or RCBO. Each appliance connected to the circuit is protected locally by a fuse of the appropriate rating for the appliance wattage contained either within its plug or in the FCU to which it is wired.

Old radial Circuits

Before the introduction of the ring circuit, socket outlets were supplied by individual circuit cables running from a fuseway in the house fusebox. The outlets themselves had round holes designed to accept round-pin plugs; both were made in three different sizes. The small 2-amp plug and socket was designed to supply table or standard lamps and small appliances such as radios. The medium-sized 5-amp type supplied larger appliances without heating elements – vacuum cleaners, for example. The largest 15-amp size catered for powerful appliances such as electric fires and kettles. The plugs were not fused, and protection for the appliance and its supply cable was provided solely by the rewirable circuit fuse in the fusebox. Each circuit was supposed to feed just one socket, but extra sockets were often added to increase the capacity of the system (but this also increased the risk of overloading it).

MODERN RADIAL CIRCUITS

Modern radial circuits supplying 13-amp socket outlets are quite different from their forebears. They are wired up in the same way as lighting circuits, with the circuit cable running from the fuseway in the consumer unit to the first socket outlet and then on to the others in series, terminating at the most remote socket. They are mainly used in two particular situations; to supply long buildings where completing a ring circuit could effectively double the amount of cable needed, and to supply home extensions too big to be fed from an existing ring circuit without exceeding the floor area limits imposed by the Wiring Regulations.

Radial circuits can be wired up in one of two ways. If 2.5mm² cable is used, the circuit must be protected by a 20-amp fuse or MCB and can serve rooms with a floor area of up to 20sq m (215sq ft). For larger floor areas up to a maximum of 50sq m (540sq ft), the circuit should be run in 4mm² cable and the circuit must be protected by either a 30- or 32-amp cartridge fuse (not a rewirable fuse) or a 30- or 32-amp MCB. Either type can feed an unlimited number of outlets.

The 30-amp radial circuit is seldom used for domestic wiring work; the high cost of 4mm² cable means it is generally more economical to use ring circuits run in 2.5mm² cable.

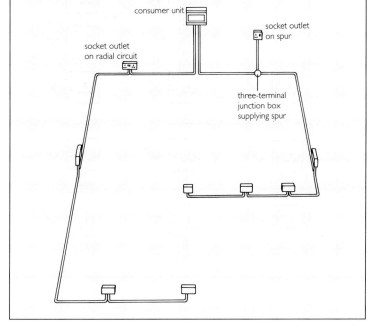

A radial circuit is wired with a single cable running from the consumer unit to each socket outlet in turn. It terminates at the most remote socket outlet. Two radial circuits are shown in this example

Below: You can only
take a spur from a
socket on the main cir-
cuit that is not already
supplying a spur
Bottom: A typical spur
providing power to a
permanently installed
appliance via a fused
connection unit

SPURS FROM POWER CIRCUITS

It is clearly impractical for the main circuit cable to run to every required socket outlet position in the house. The solution is to treat the circuit cable as the 'main line', and to add 'branch lines' - single lengths of cable called spurs - to feed outlets remote from the main circuit. These spur cables can be connected to the main circuit in one of two ways; directly into the back of a socket outlet on the main circuit cable, or by linking them to the circuit cable at a convenient point using a 30-amp three-terminal junction box. The spur cable is then run to the socket outlet position, and is connected into the socket faceplate in the usual way. As many spurs can be fitted as there are outlets or FCUs on the main circuit, but each spur may supply only one single or double socket outlet or one FCU.

SWITCHED SPURS

Where socket outlets are sited at low level in fitted kitchens to power slide-in appliances such as fridges, dishwashers and washing machines, it is impractical to have to pull the machine out to switch it off for servicing or defrosting. In this situation the socket should be wired up as a spur, with the cable running to a 20-amp double-pole (DP) isolating switch sited just above worktop level and labelled to indicate what appliance it controls. Some accessory manufacturers offer DP switches with a range of appliance names printed on the switch faceplate.

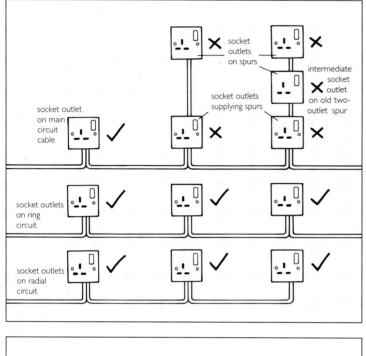

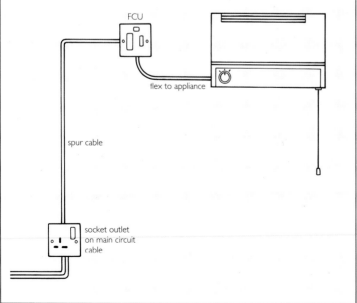

OTHER POWER CIRCUITS

Below: Wiring for a free-standing cooker

Bottom: Built-in ovens and hobs - the cables can run from one unit to the other or both from the same switch if both units are within 2m of the switch; otherwise use two switches

Apart from lighting circuits and circuits supplying socket outlets, a typical domestic wiring system will have at least two and possibly more additional power circuits supplying individual appliances that are large current-users. Each circuit is wired up as a radial circuit, with the cable running from an appropriately rated fuseway in the consumer unit to a double-pole (DP) isolating switch and then on to the appliance itself.

COOKER CIRCUITS

An electric cooker is likely to be the largest user of electricity in the house; one with two ovens, a grill and four hotplates could have a total theoretical current demand of as much as 50 or 60 amps, although seldom will every heating element be on at the same time. To take account of this actual usage, a principle called *diversity* is used to assess the rating of the protective device needed on the cooker circuit.

To do this, start by finding out the wattage rating of the cooker, which is usually given on a plate on the back of the appliance. Divide this by the mains voltage (240 in most areas) to get the maximum current demand. For a cooker rated at 12kW this is 12000 divided by 240 = 50 amps. The diversity principle assumes that the first 10 amps of demand are always needed, plus 30 per cent of the remainder. So the current demand will be 10 amps plus 30 per cent of 40 amps (12 amps), giving a total demand of 22 amps. If the circuit supplies a cooker control unit containing a 13-amp socket outlet, an additional demand of 5 amps is added for the socket, giving a total of 27 amps. A 30-amp fuse or MCB would be the appropriate protective device for this circuit.

The next task is to size the circuit cable, and this requires even more complex calculations. In practice, 6mm² cable is normally used for cooker circuits so that even if the cooker in use does not require this size of cable, any upgrade in the future will not call for a rewire.

For a free-standing cooker, the circuit cable is run to an appropriately rated **cooker switch** or a **cooker control unit**, which is sited to the side of the cooker position and about 300mm above the worktop level. From there a further length of cable is run down to a point in the centre of the cooker space, about 600mm above floor level, where it is connected to a

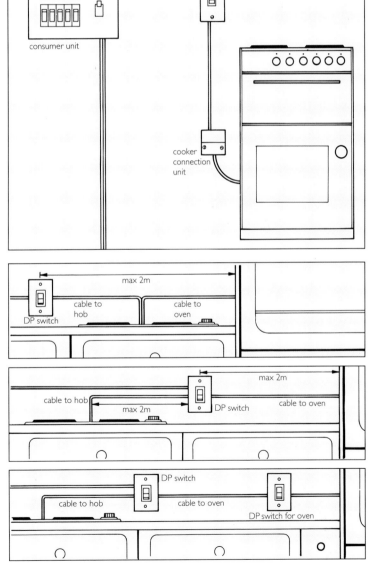

consumer unit

double-pole cooker switch

cooker connection unit

max 2m

DP switch

cable to hob

cable to oven

cable to hob

max 2m

max 2m

DP switch

cable to oven

DP switch

cable to hob

cable to oven

DP switch for oven

cooker connection unit (also called a cooker outlet plate). The cooker is wired to this using cable of the same rating as the circuit cable, and long enough to allow the cooker to be pulled out of its recess. This is the only situation where cable rather than flex is used to connect a movable appliance to the mains.

Separate built-in ovens and hobs can be controlled from one cooker switch so long as both components are within 2m of the switch. The incoming circuit cable runs to the switch; then either one cable is looped first to one component, then on to the other, or a separate cable is taken from the switch to each component. If one component is more than 2m from the main switch, a second isolating switch must be wired in to the cable supplying the second component within 2m of it. The cable for both cooking appliances should be the same size as the circuit cable.

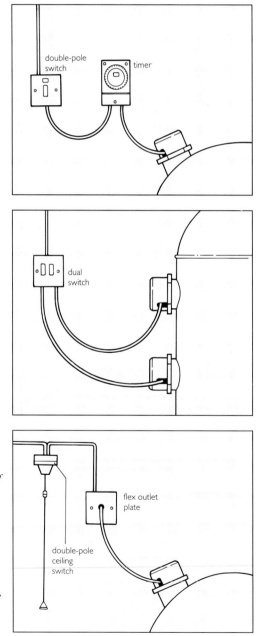

An immersion heater has its own circuit, running to a double-pole switch close to the heater position. The heater is linked to the switch with heat-resistant flex. Top: A timer can be wired in if required. Centre: A special dual switch can be used to control twin or dual elements. Bottom: If the switch is ceiling-mounted the cable is taken on to a flex outlet plate and the heater is connected to this

IMMERSION HEATER CIRCUITS

An immersion heater is widely used as a back-up to hot water systems that are centrally heated, and in this situation it is wired up on its own circuit from a 15- or 16-amp fuseway in the consumer unit. The circuit is wired in 2.5mm² cable, which is normally run to a 20-amp double-pole (DP) isolating switch with flex outlet and neon indicator located close to the hot water cylinder. The connection between the switch and the heater thermostat is made with a length of 1.5mm² three-core heat-resisting flex (reference 3183TQ). This may run via a clock timer if one is required to control the water heating periods. If a twin-element heater or dual elements are fitted, they can be controlled by a special dual switch; see page 105 for details.

If the hot cylinder is located in a bathroom, the switch controlling the heater must be placed out of reach of anyone using the bath or shower if it is within the room. The assessment is made as though there were no door on the airing cupboard. Alternatively a ceiling-mounted cord-operated switch is fitted within the room.

Remote switching of the heater, from the kitchen for example, is an additional convenience. In this situation the circuit cable runs first to a 20-amp DP switch with neon indicator in the kitchen, and then on to another switch next to the hot cylinder. The link to the heater is made with 1.5mm² heat-resisting flex as described above. The same arrangement is used to wire off-peak immersion heaters.

WATER HEATER CIRCUITS

Small instantaneous water heaters – the type often installed to serve washbasins remote from the house plumbing pipework – can be supplied from a socket circuit so long as they are rated at no more than 1.5kW. The connection should be via a switched FCU fitted with a 13-amp fuse. This must be sited so there is no chance it will be splashed with water. Larger heaters, including under-sink and multi-point types, must be supplied by an individual circuit run from separate fuseways in the consumer unit. These are usually wired using 2.5mm² cable, a 20-amp DP isolating switch and a 20-amp circuit fuse or MCB, but higher-rated cable and switchgear may be needed for some installations.

SHOWER CIRCUITS

Electric showers are relatively inexpensive to buy and easy to install when compared with a conventional shower fed from the hot and cold water supplies. They are big current users, however, and must have their own circuits. In properties with a 60-amp service fuse and a cooker circuit in use, expert advice should be sought before an electric shower is installed, to avoid possible overloading of the system.

How the circuit is wired depends on the current demand of the shower heater. Use 6mm² cable for shower heaters up to 9.2kW, so that even if the unit being wired up is of a much lower rating there will be no need to rewire should you upgrade it. Larger cable may be required for long circuits. For units up to 7kW use a 30- or 32-amp protective device; for units between 7kW and 9kW use a 40-amp device.

The circuit cable is run from the fuseway to a 40- or 45-amp ceiling-mounted cord-operated DP isolating switch near to the shower cubicle. This switch should have a neon indicator light, and must also have a mechanical on-off indicator flag so that the user can see whether the unit is switched on even if the neon has failed. Cable is then run on from the switch direct to the shower heater terminals.

It is desirable for safety reasons that shower circuits have RCD protection. This can be provided either by including an RCD in the circuit wiring between the consumer unit and the DP switch, by placing the shower fuseway under RCD control in a split-load consumer unit or by protecting the shower circuit with an RCBO. The RCD should be a 30mA type.

Finally, it is essential that the shower supply pipework is independently cross-bonded to earth. See pages 66 and 108–9 for more details.

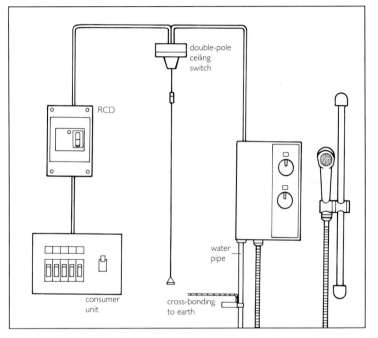

A shower circuit runs from its own fuseway in the consumer unit to a double-pole switch and then on to the shower unit itself. It should be RCD-protected

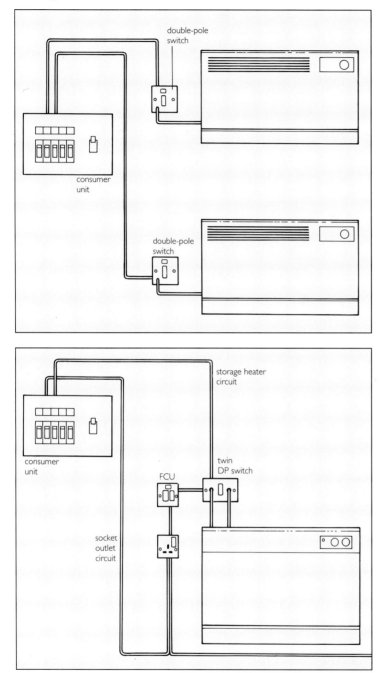

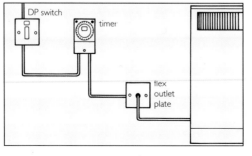

Top: Storage heaters each have their own radial circuit, running to a double-pole switch with flex outlet close to the heater. Centre: If the heater is fan-assisted a separate supply for the fan is wired from a socket circuit

STORAGE HEATER CIRCUITS

Two types of off-peak storage heaters are in common use today; the fan-assisted type, and the combined storage/convector. Both are charged up at cheaper night-time rates (you need a dual-rate meter for this), and discharge their heat during the day. The fan-assisted type incorporates a fan to boost the daytime output rate of the heater. The combined type contains an electric convector heater rated at 1, 1.5 or 2kW which provides additional heating (at full price) during the day if the storage capacity of the heater has been exhausted.

Both types have their own separate radial circuit, run from fuseways in a separate off-peak consumer unit. In the past circuits traditionally fed two heaters, but to comply with the Wiring Regulations each circuit must now supply only one. The circuit cable runs to a double-pole (DP) isolating switch fitted close to the heater position. The heater is then connected to the switch with a length of heat-resisting flex. For heaters rated at up to 4kW, the circuit is wired using 2.5mm² cable, with a 20-amp DP switch and a 20-amp fuse or MCB acting as the protective device. For more powerful heaters 6mm² cable is used, plus a 30/32-amp DP switch and a 30/32-amp fuse or MCB.

With fan-assisted types, the fan is obviously required during the day so its power supply must be taken from the 24-hour supply. The usual way of doing this is via a switched fused connection unit (FCU) fitted with a 5-amp fuse and run as a spur from a socket circuit. The FCU should be fitted next to the heater's DP switch, so that the flex leading to the fan can be connected easily. Alternatively, a special twin DP switch rated at 25 amps and with two flex outlets can be used to control both heater and fan. The single on-off switch isolates both heater and fan. The fan is fed via an unswitched FCU.

With combined storage/convector types, the convector must also be connected to the 24-hour supply. This is best done via a switched FCU fitted with a 13-amp fuse and again run as a spur from a socket circuit.

Electricity supply companies are currently introducing changes to the cheap-rate supplies they offer. Check with yours first before installing storage heaters.

Left: An in-line timer can be included

OUTDOOR CIRCUITS

Electricity is just as useful out of doors as it is within the house. Lights can illuminate the approach to the house after dark, help deter intruders and also let you enjoy the patio and garden on summer evenings after the sun has gone down. A power supply means you can use all sorts of gardening and leisure equipment without the need for long extension cords run out from the house, and having electricity laid on in garages, garden sheds and greenhouses is an obvious boon for car mechanics, do-it-yourselfers, craft enthusiasts and gardeners alike.

It must not be forgotten, however, that 'electricity' and 'outdoors' are potentially a very dangerous combination. For a start, everything must be carefully protected to keep water out of the wiring. Even more important is the fact that you are outside the area protected by the house's earthing system, so the greatest care must be taken to guard against the risk of receiving an electric shock. This means having double-pole switch isolation and RCD protection on all outdoor circuits, and whenever possible using only electrical equipment that is double-insulated.

CIRCUITS TO OUTSIDE LIGHTS

Lights on the external walls of the house can be wired up as extensions of the indoor circuits as long as the house circuit will disconnect within 0.4 seconds in the event of an earth fault. Spurs can be taken from either a lighting or a socket circuit as appropriate in the same way as wiring up an additional light fitting indoors, but check that the indoor circuit is not already fully loaded before adding extra lights. This makes it a simple task to fit lights next to all exit doors or to illuminate drives and patios. However, the fittings used must be weatherproof and the wiring to them must be sealed to keep water out. Unless it runs directly into the back of the fitting through the wall, it should be protected from damage by being run in rigid sealed PVC conduit clipped to the house wall.

Lights remote from the house must be run as separate circuits originating at a 5/6-amp fuseway in the consumer unit. The cable can be run above ground, either overhead via a series of individual posts or along walls (but **never** along fences, which can blow down and sever the cable). Alternatively it can be run underground, so long as it is buried to a depth at which it will not be damaged – at least 450mm (18in) below ground level. See *Circuits to outbuildings* for more details. The circuit cable can run directly into the individual light fittings via weatherproof seals, or can supply a series of outdoor socket outlets (see page 51), allowing

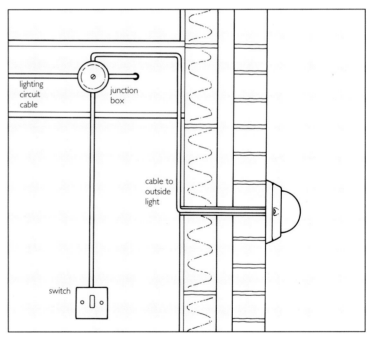

Lights mounted on the house wall can be wired as spurs from the house circuits. This page: Outside light fed from indoor lighting circuit. Next page, top: Outside light fed from socket outlet circuit

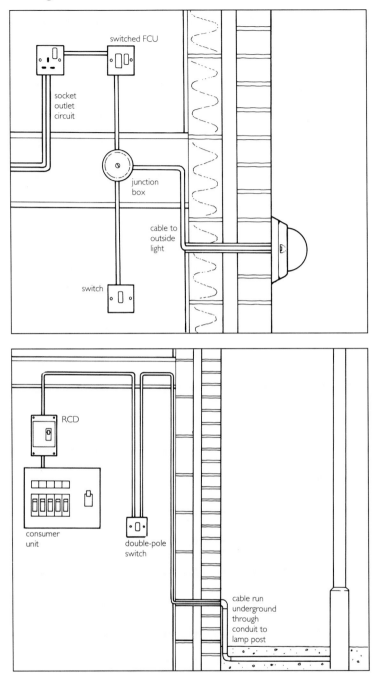

individual lights to be moved around as required. The entire circuit must be capable of being isolated from the mains with a double-pole switch, and should have RCD protection too (this is mandatory for outdoor socket outlets).

An alternative to a series of outdoor light fittings is to use **festoon lighting** – essentially a length of heavy-duty outdoor-quality flex to which a series of weatherproof lampholders are permanently attached. These either take standard bayonet-cap lamps or Edison screw types, and are available in a range of lengths and lamp spacings. Festoon lighting is designed to be draped round trees or suspended from garden structures or catenary wires, and can be left in position permanently so long as care is taken to prevent high winds from dashing the lamps against walls, branches and so on. Wire guards or pop-on plastic lanterns can help prevent accidental damage to the lamps. Festoons should always be supplied from a weatherproof outdoor socket outlet which has RCD protection.

EXTRA-LOW-VOLTAGE LIGHTING CIRCUITS

The safest type of outside lighting is extra-low-voltage (12-volt) lighting, supplied from a small weatherproof safety isolating transformer made to BS3535 located under cover. The extra-low-voltage cable can be safely run on the surface of flower beds or can be clipped to walls or fences, and the lights are connected to it via a bi-pin fitting that pierces the cable insulation to make the electrical contact. The transformer can usually supply up to four lights, and is itself simply plugged into a suitable socket outlet. If the transformer is located outside the area protected by the house's earthing system, the socket must have RCD protection since it could also power other electrical equipment being used out of doors.

Top: Outside light mounted on house wall fed via a
fused spur from a socket outlet circuit

Bottom: Outside lights remote from the house must be
wired on their own circuit running from a fuseway
on the consumer unit. A double-pole switch provides
complete isolation from the mains and RCD protection
gives additional safety

OUTDOOR SOCKET CIRCUITS

As garden power tools become ever more popular, there is a growing need for the convenience of having one or more special socket outlets out of doors. An extension lead taken from an indoor socket outlet has two main drawbacks; a long flex is prone to accidental damage and may suffer from voltage drop, and the socket outlet used may not have RCD protection on an older wiring installation. The latter drawback can be remedied by using an RCD plug or adaptor, but there is really no substitute for having dedicated socket outlets. Not only will they supply d-i-y and garden power tools; they also enable you to use other appliances out of doors – a vacuum cleaner to clean out the car, for example, or the coffee machine and toaster for a patio breakfast – and to plug in outside lights at convenient points.

A single socket outlet on the house wall can be wired up as a spur from one of the house's socket circuits. It must have RCD protection, and the best solution is to fit a special weatherproof socket outlet containing its own RCD. One-and two-gang outlets are available, both fitted with a 30mA RCD.

Socket outlets remote from the house must be supplied via a separate RCD-protected circuit run from a 30/32-amp fuseway in the consumer unit or in a separate switchfuse unit. The RCD must be a 30mA type, and the circuit should also be capable of complete isolation from the mains, using a suitable double-pole switch.

The cable can be run overhead, but is better run underground to the various socket outlet positions. The socket outlets themselves can be mounted on walls or sturdy posts, and the cable or conduit run should be taken right up to the outlet via a waterproof cable entry seal. See *Circuits to outbuildings* over the page for more details.

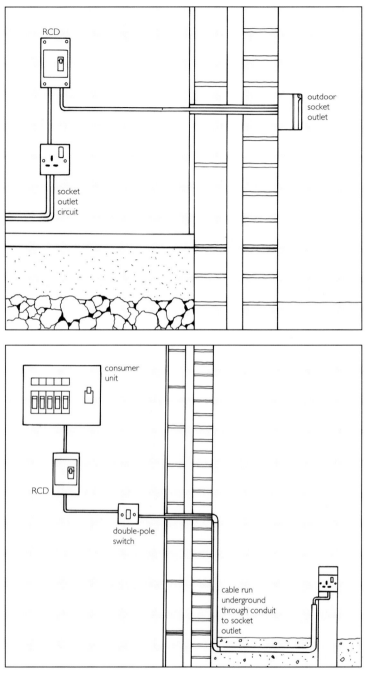

Top: An outdoor socket outlet mounted on the house wall can be wired as a spur from an indoor socket outlet circuit. It must be RCD-protected

Bottom: Outdoor socket outlets remote from the house must have their own circuit run from the consumer unit and capable of isolation by means of a double-pole switch. The whole circuit must be RCD-protected

CIRCUITS TO OUTBUILDINGS

Many outbuildings such as detached garages, garden sheds, greenhouses and summerhouses will be much more versatile if they have their own power supply. As with circuits to outside lights and socket outlets, the circuit can be taken overhead or underground. For a typical-sized garden, 2.5mm² cable can be used to supply a 20-amp radial circuit and 4mm² cable for a 30-amp circuit. Use cable one size larger on long cable runs.

Overhead wiring is undoubtedly the easier of the two to install, although it is ugly and is prone to accidental damage by carelessly wielded ladders or by high winds. Ordinary two-core-and-earth PVC-sheathed cable can be used for the whole cable run, and needs no support for spans of up to 3m. Spans up to this length can be given additional protection by being run in a length of 20mm heavy-gauge galvanized steel conduit with no joints in the span, suitably secured and earthed and fitted with protective plastic bushes at each end to prevent the cable sheath from chafing. Longer spans must be supported on a catenary wire, which must itself be earthed. The span must have at least 3.5m clearance above ground level, increased to 5.2m if it crosses areas accessible to vehicles, such as a driveway.

Underground wiring involves more work initially, but once installed the circuit cable is safely out of harm's way. The underground section of the cable run can be wired in ordinary two-core-and-earth PVC-sheathed cable if it is protected along its entire length with rigid PVC conduit connected with solvent-welded couplers and fittings or can be run in armoured cable. The cable run should be buried to a depth where it cannot be damaged: about 450mm, and deeper – up to 700mm – under garden areas likely to be double-dug. It must be covered with suitable coloured tape or markers so that anyone digging will become aware of the presence of cables before severing them. It can be additionally protected by the use of rigid cable covers.

The underground conduit run must be fully assembled before the cable is drawn into it. At the house and outbuilding ends of the run, 90° bends are used to take the conduit up the wall of the building to a convenient entry point, which should be sealed with mastic to prevent water penetration. The conduit should pass right through the wall with a slight downward slope to the outside, to protect the cable from chafing.

Within the house, the cable runs to a separate 20- or 30-amp fuseway in the existing consumer unit or in a new small consumer unit if no spare ways are available. The circuit must be protected by a 30mA RCD.

Within the outbuilding, the cable should run first to a double-pole isolating switch and then continue as a radial circuit to a series of socket outlets, with lights being fed via a fused connection unit. Alternatively, the incoming cable can be taken to a two-way consumer unit fitted with an isolating switch, a 20-amp fuse or MCB to supply socket outlets and a separate 5-amp fuse/MCB for lights.

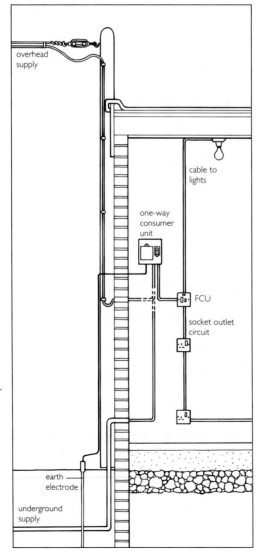

The power supply to an outbuilding can be run overhead or underground. Within the building the cable runs to either a double-pole switch or to a one-way consumer unit with an isolator switch (as illustrated). The cable is then taken on to supply socket outlets. Power for lighting is taken as a spur from a fused connection unit. Alternatively separate socket outlet and lighting circuits could be run from a two-way consumer unit. The outbuilding is earthed via its own electrode

overhead supply

cable to lights

one-way consumer unit

FCU

socket outlet circuit

earth electrode

underground supply

WIRING TECHNIQUES

WORKING WITH FLEX AND CABLE
MOUNTING WIRING ACCESSORIES

WORKING WITH FLEX & CABLE

Flex and cable are the veins and arteries of your house wiring system, carrying electricity from its heart – the consumer unit or fusebox – to wherever it is needed. It is essential for the system to have good, healthy circulation, which means that all the flex and cable used throughout the system must be in good condition and that all connections made between them and other

components of the system must be perfect. Defects do not just make the system inefficient; they can make it downright dangerous. Follow this chapter to ensure that you prepare flex and cable correctly for use, make the connections between the current-carrying conductors properly and safely, and run the circuit cables around the house in a suitable manner.

Remember that **flex** is used solely for connecting electrical appliances to the mains via plugs and socket outlets or via fused connection units, and for linking pendant lampholders to their ceiling roses. It is never used for any circuit wiring: that is the job of **cable**, which is also used to link freestanding cookers to the mains (there is no flex big enough to carry the current demand).

PREPARING FLEX
FOR CONNECTION

Both flex and cable must have some of their outer sheathing and core insulation removed to allow their conductors to be connected to the terminals of the system's various components. Since the insulation is there to prevent the conductors from coming into contact with anything other than the terminals to which they are connected, it is essential to remove no more of it than necessary. Similarly, the sheathing is there to protect the conductors from damage, so just enough should be removed to allow the cores to reach the terminals inside the wiring accessory or plug, while ensuring that the sheathing terminates within it.

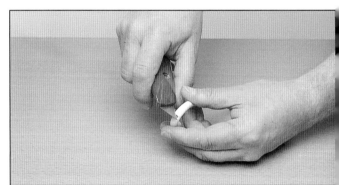

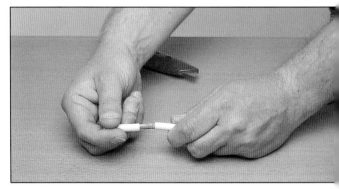

THINGS YOU NEED
- Appropriate flex
- Side cutters
- Handyman's knife
- Wire strippers
- PVC insulating tape

1 Check that you are using the right type of flex for the job, and that it is of the correct current rating – see page 23. Cut the length you require for the job using side cutters.

2 With PVC-sheathed flex, mark the sheathing at the point where you want to cut it. Then fold it over at that point and nick the sheathing with your knife. The tension at the bend will open up the cut to about one-third of the flex circumference. Fold it the other way and nick it again, repeating the operation a third time if necessary to extend the cut all the way round the sheathing. This technique avoids any risk of cutting into the core insulation, which must be avoided at all costs. Pull off the unwanted section of the sheathing and discard it.

3 With fabric-covered non-kink flex, cut carefully through the fabric braid first and remove it, then cut away the rubber sheathing underneath as in Step 2. Wrap the cut braid in PVC insulating tape to stop it from fraying.

4 With flat two-core flex, used on some double-insulated appliances, either use the same technique as in Step 2 or slit the sheathing lengthways so the tip of the knife blade runs between the two cores. Peel the sheathing back and trim off the excess; this is best done with side cutters. Check that you have not sliced into the core insulation; if you have, cut off the exposed cores and start again.

5 Check how long the cores need to be to reach their terminals, and cut them to length with your side cutters. Then use your wire strippers to remove about 12mm (½in) of the insulation from each core. Most wire strippers can be adjusted – or have a range of jaw sizes – to suit the different core diameters you are likely to encounter. Check that you select the right setting by testing it on a core offcut first, to ensure that you do not cut through any of the fine strands that make up the conductor. If you do this, you reduce the current-carrying capacity of the flex, which could cause overheating at the terminal. Cut off any cores you have nicked in this way and start again.

6 Twist the strands of each core together neatly with your fingers, ready for connection to their terminals.

7 When new appliances come without a plug, the flex cores are often already stripped, with the core strands soldered at the ends. However, the cores are rarely cut to the right length and the solder should not be used to make the contact with the plug terminals, so cut them off flush with the end of the sheathing and prepare the flex from scratch as described above.

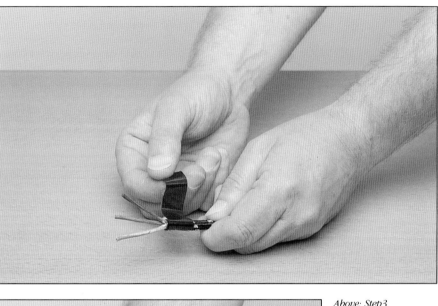

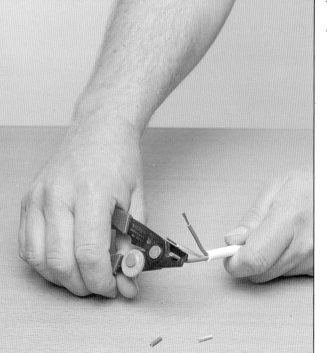

Above: Step 3

Left: Step 5

WIRING A PLUG

Plugs provide the physical connection between the flex of an electrical appliance and the mains. Most homes now have socket outlets that accept fused 13-amp plugs with three rectangular pins, although older systems may still have outlets with round holes that take round-pin plugs.

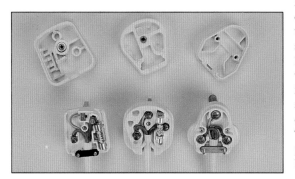

Until recently, few appliances were sold with plugs attached; it is now usual for all new electrical equipment to be sold with a plug fitted, but you will still need to know how to fit a plug to existing appliances or to replace one.

All fused plugs now on sale must have the live and neutral pins sleeved in plastic to reduce the risk of fingers – especially children's – touching live metal as the plug is withdrawn from its socket. Check that any fused plugs you buy are also marked 'made to BS1363' (BS1363A for plugs with resilient casings, or BS546 for unfused round-pin plugs) and carry the BS Kitemark, or else have ASTA certification. There are plenty of cheap plugs on the market, sold in bargain stores, street markets

and the like, that do not meet the British Standard or ASTA requirements, and which could be unsafe to use. If you find any, notify your local authority Trading Standards Officer.

The rating of the plug fuse must match the current demand of the appliance – 3 amps for those rated at up to 720W, 13 amps otherwise (some TVs need a 5-amp fuse). While changing the fuse is a simple matter, you may prefer to buy plugs already fitted with the correct fuse. Most retailers now stock plugs fitted with both 3-amp and 13-amp fuses.

Left to right: 13-amp plug with bar-type cord grip and pillar terminals, 13-amp plug with jaw-type cord grip and stud terminals, round-pin plug

THINGS YOU NEED

- Plug with appropriate fuse
- Medium screwdriver
- Terminal screwdriver
- Appropriate flex
- Side cutters
- Handyman's knife
- Wire strippers
- PVC insulating tape

1 Start by opening the plug, setting its screw aside unless it is captive. Then prepare the flex (page 54). Hold the flex over the open plug so you can see how long the longest core must be – it is usually the earth core, although on some plugs all three cores are the same length – and mark and remove the sheathing.

2 Lay the flex over the plug with the sheathing over the cord grip, and position each core in the channel that leads to its terminal – brown (live) to the terminal marked L, blue (neutral) to the one marked N and green/yellow (earth) to the one marked E or ⏚.

Use this simple reminder to make sure you get it right. With the plug open and facing

you with the flex inlet at the bottom, connect the **BR**own core to the **B**ottom **R**ight terminal and the **BL**ue core to the **B**ottom **L**eft terminal. The earth core goes to the top terminal (not used with two-core flex on double-insulated appliances).

3 If necessary, cut each core down in length. The live and neutral cores should reach just past the top of the terminal, while the earth core should have a little more slack on it for safety's sake: if the flex is yanked and the cord grip does not do its job properly, the live or neutral core will be pulled out or broken before the earth core.

Step 1

4 Strip the core insulation to expose the conductor strands. Remove about 6mm (¼in) of insulation for connection to pillar terminals, and about 13mm (½in) for stud terminals. Twist the strands together neatly.

5 On some plugs, you may have to remove the fuse first to get at the live terminal; simply prise it out with a screwdriver. On others, the pins may be loose, which may make it easier to insert the cores in the terminals, but which also provides three more components to drop.

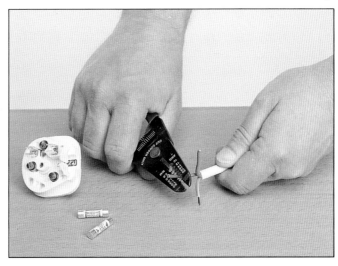

Step 4

Safety tip

If you have to replace a moulded-on plug, cut it off, hammer the pins out of line so it cannot be plugged into a socket outlet, and throw it away.

6 With pillar terminals, loosen the terminal screw and push the core into the terminal hole until the insulation just touches the side of the pin. Tighten the screw down fully. Repeat for the other terminals.

7 With stud terminals, remove the studs and wind the cores clockwise round the pins. Screw the studs back on by hand, checking that there are no stray conductor strands, and tighten them down with a screwdriver.

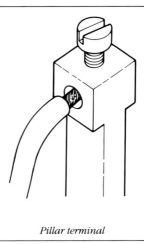

Pillar terminal

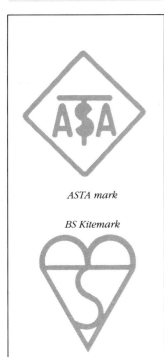

ASTA mark

BS Kitemark

8 Check that each core lies in its channel in the plug body, with no kinks or sharp bends, and that it will not be trapped when the plug is closed. Then secure the flex sheathing in the cord grip, either by pressing it into nylon jaws or by undoing the screws of a bar-type grip and then tightening it down at both sides to clamp the sheathing securely. Test the cord grip's efficiency by giving the flex a sharp tug.

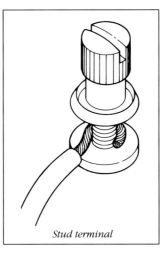

Stud terminal

9 Double-check that each core has been connected to the right terminal, that no conductor strands are visible, that all terminal screws are tight and that the correct fuse has been fitted. Then fit the plug cover.

PLANNING CABLE RUNS

If you are planning to extend or alter your house wiring, you have to work out how you are going to run the cable from its connection point on the existing system to its final destination – a socket outlet wired up as a spur from an existing socket circuit, for example. Ideally cable runs should be concealed, as they are when a new house is wired up; it is not difficult to do this, although it will involve some disruption.

Most cables are run in ceiling and floor voids, dropping down or rising up as necessary to serve wall-mounted wiring accessories such as light switches and socket outlets. How you conceal new cables depends on which circuit you are working on, and on the type and structure of your house. In a typical two-storey house with a loft, access to the lighting circuits serving upstairs rooms is obviously easy, and in an older home there may be a substantial crawl space beneath the ground floor allowing access to the downstairs socket and power circuits. In houses with solid ground floors, the downstairs power circuits have to be fed by cable drops from the floor above. Everything else will have to be run in the void between the downstairs ceilings and the upstairs floors, and access can be gained only by lifting floorboards, which means moving fitted furniture and carpets. Finally, vertical cable runs are either run in conduit and concealed in channels (called chases) cut in the plaster, or are hidden within stud partition walls. Some typical cable routes are shown in the illustration.

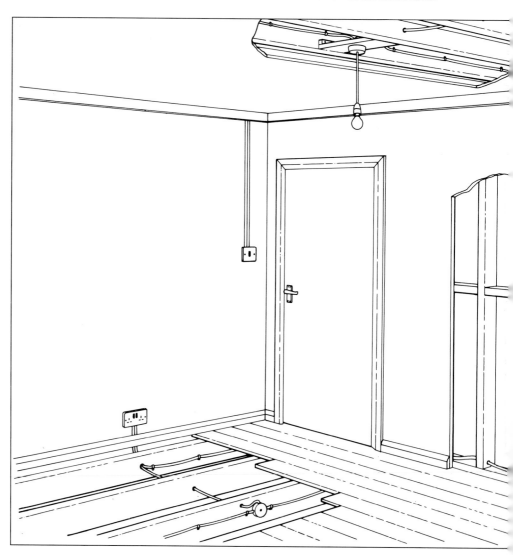

Cable is mainly run in ceiling and floor voids, and reaches wall-mounted wiring accessories either in conduit buried in chases cut in plaster or by being run within stud partition walls

Concealing new cable runs can mean major disturbance of existing decorations and, for this reason, you may prefer to consider some form of surface wiring as a short-term solution, leaving the cutting of chases and the lifting of floorboards necessary for complete concealment until the room is next stripped for redecoration.

When planning possible cable runs across wall surfaces, remember that cables should ideally run vertically at all times. Horizontal runs are permissible, but diagonal ones are allowed only if they are run in earthed steel conduit.

Since PVC-sheathed cable is a flattened oval in cross-section, it will lie flat against wall and ceiling surfaces so long as it is secured at intervals (see below). It can be formed into gentle curves in the plane of the oval, and into sharper bends at right angles to it – to avoid kinking the cable the radius of the bend should be at least four times the cable width – allowing the cable to be taken round internal and external corners and to change direction.

On wall surfaces, it is best to make use of the natural protection offered by mouldings such as skirting boards and door architraves, and to fix the cables to the wall immediately above the top edge of the former and close to the square edge of the latter. Cable should not be fixed to the face of these mouldings, however temptingly easy this may be to do, since it could be damaged by carelessly moved furniture. Cable can also be run round the room walls just below ceiling level, and in room corners. On ceilings, the cable can obviously run in any direction across the surface.

Cable clips are made to suit the various commonly used cable sizes, including some designed to hold two cables side by side. They are usually fitted with a small masonry pin which will provide a firm fixing in plaster or plasterboard and which can also be driven into wood. They should be fitted at the spacings given for horizontal and vertical cable runs in the table below, with extra clips being used to secure the cable at each side of any bends.

SURFACE-MOUNTING CABLE

The simplest and quickest way of running a new cable between two points is to take it across wall and ceiling surfaces. Modern PVC-sheathed cable is tough enough to be exposed in this way, and the only drawback is that it looks rather untidy. The cable is fixed in place at intervals with small moulded plastic cable clips.

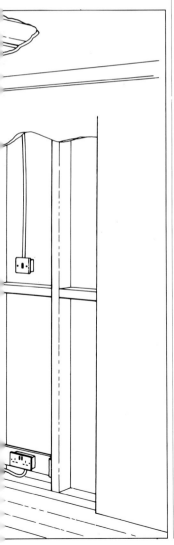

THINGS YOU NEED
- Cable clips to match cable size
- Pin hammer
- Tape measure
- Side cutters for cutting cable

Cable Clip Spacings

Cable size	Horizontal runs	Vertical runs
1mm²	250mm	400mm
2.5mm²	300mm	400mm
4/6mm²	350mm	500mm

USING MINI-TRUNKING

Mini-trunking is an improvement on straightforward surface-mounting, since the cable is both concealed and protected by the neat plastic moulding with its snap-on cover. The trunking can be run across both wall and ceiling surfaces as required.

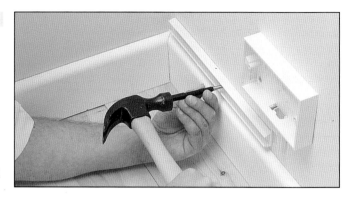

THINGS YOU NEED

- Mini-trunking
- Connectors, corner and T-pieces
- Mounting boxes and adaptors
- Tape measure
- Pencil
- Fine-toothed saw
- Fixing pins (unless you are using self-adhesive trunking)
- Hammer
- Nail punch

Mini-trunking is available in a range of sizes. The smallest has a 16mm (⅝in) square cross-section, ideal for a single cable run, while larger sizes will accommodate several cables side by side. Note that if a number of cables are run together you should check with a professional electrician whether you need to use larger cable sizes for safe operating of the system. All sizes of mini-trunking usually come in 3m (10ft) lengths. There is a full range of connectors, internal and external angle mouldings, T-shapes and stopends, making it an easy job to construct any cable run required. There are even special surface mounting boxes and adaptors made to match the mini-trunking in most ranges.

You can also buy larger types of trunking which are designed to replace existing skirting boards, architraves and cornices.

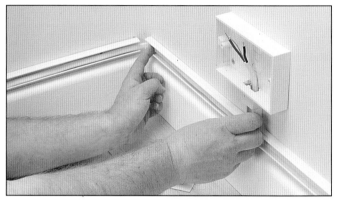

I Fit the accessory mounting box complete with its outlet adaptor to the wall first (see page 69). Then measure up and cut the mini-trunking to length, making allowances for the various fittings along the run. Unclip the cover strip and pin the base section to the wall with small masonry pins, using a nail punch to drive the

pins home without damaging the moulding. If you are fitting self-adhesive mini-trunking, simply peel off the release paper and press it into place.

2 Position the cable in the mini-trunking, forming bends as necessary and feeding the cable end into the mounting box via the adaptor and knockout.

3 Snap the cover strips on to the base sections, aligning their ends carefully.

4 Press the connectors, T-pieces and corner pieces into place at all the joints to complete the installation.

From top: Steps 1, 2, 3 and 4 (left to right)

Cable in a chase is run in lengths of oval PVC conduit, held in place with galvanized nails or masonry nails and plastered over. The conduit provides some additional protection for the cable, and it allows the cable to be removed and replaced easily if this become necessary in the future. Use of conduit has only recently become mandatory, so old chases may contain uncovered cables held in place with cable clips.

Oval conduit is available in a range of sizes from 13 x 8mm (½ x ⅜in) up to 29 x 11mm (1⅛ x ⁷⁄₁₆in), in standard 3m (10ft) lengths. Solvent-weld couplings are used where necessary to join lengths.

I Use your plumb line, straightedge and pencil to mark out the route for the chase on the wall surface. Make the chase about 6mm (¼in) wider than the conduit.

2 Cut along both sides of the chase with the brick bolster and club hammer until you reach the masonry beneath. The plaster between the parallel cuts may simply fall out as you work; otherwise lever or rake it out as you go to leave as neat a channel as you can. Wear work gloves unless the brick bolster has a hand guard.

3 If you have a lot of chases to cut it is worth hiring a special power tool called a chasing machine. This has two adjustable diamond-tipped cutting wheels, and will cut perfect chases up to 20mm (¾in) deep and 23mm (⅞in) wide. It creates huge amounts of dust, and is best used connected to a dust extraction unit. Always wear safety goggles and a face mask when using one.

4 Secure the conduit in the chase with galvanized plasterboard nails or small masonry nails driven in each side.

5 Feed cable into the conduit and plaster over the chase. Make sure you leave enough cable protruding from the mounting box at the end of the chase to allow the wiring accessory to be connected easily.

RUNNING CABLE IN SOLID WALLS

Cable runs to wall-mounted accessories such as light switches and socket outlets are concealed in vertical channels (chases) cut in the plaster. The cable can then enter either a flush or a surface mounting box.

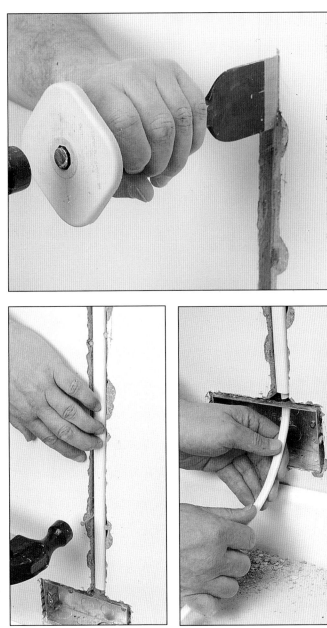

THINGS YOU NEED

- Plumb bob and line
- Straightedge
- Pencil
- Sharp brick bolster and club hammer or hired chasing machine
- Oval PVC conduit with couplings
- Galvanized nails or masonry nails
- Hammer
- Patching plaster
- Filling knife or plastering trowel

Top: Step 2,
Far left: Step 4,
Left: Step 5

RUNNING CABLE IN STUD WALLS

Many homes have internal walls built as timber stud partitions – a timber frame clad with plasterboard in modern homes, or with lath-and-plaster in older ones. Hollow walls of this type provide an ideal place to conceal cables during construction, and it is fairly easy to gain access to the hollow void within an existing the wall to run new cables.

Warning

In modern timber-framed homes the inner leaf of the external cavity walls is of similar construction to a stud partition wall, although the outer face is clad with plywood sheathing and waterproof building paper rather than plasterboard. However, since the wall also contains thermal insulation and a vapour barrier that should not be pierced, no attempt should be made to run new cables within it.

Cable can be fed into a stud partition wall from above or below. Cut windows in the plasterboard to take cavity mounting boxes (see page 71) for wiring accessories, and to allow the cable to pass any noggings you encounter. If you are building a new stud wall, cables can also be run horizontally through holes drilled in the studs. There should be at least 50mm (2in) between the hole and the wall surface

THINGS YOU NEED

- Electric drill plus wood bits
- Plumb bob and line
- Straightedge
- Handyman's knife
- Hammer
- Wood chisel
- Plasterboard patches
- Plasterboard nails
- Patching plaster or filler
- Filling knife

'Paramount' Partitions

Some stud walls have what is termed 'egg crate filling'. The only way to run cable down the inside of these walls, properly known as Paramount partitions, is to punch a hole through the filling from above using length of steel conduit.

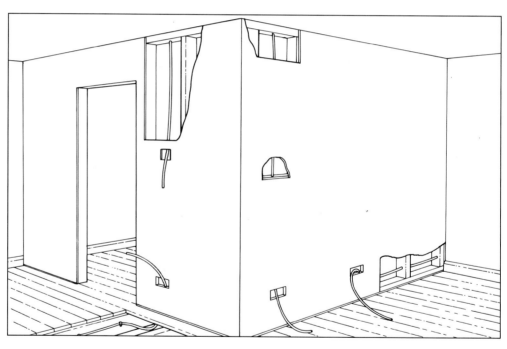

How easy it is to feed cable inside a stud partition wall depends on how the wall framework has been built, and whether you have relatively easy access to the head and sole plates (the top and bottom horizontals). Cable can be fed into the void from above via a hole drilled in the head plate, and from below via a hole in the sole plate; alternatively it may be fed up behind a skirting board if this is a simpler option.

If cables are being fed into the wall from above, it is best to use a plumb bob and line first. The bob will drop within the wall and a hole can then be made in the wall surface, allowing the line to be drawn through with the cable attached to its other end.

The two obstacles that may impede your progress are horizontal braces called noggings, included to stiffen the wall framework, and insulation blanket included to reduce

sound transmission through the wall. If the former are encountered, you will have to locate the nogging, remove a small square of plasterboard from the wall just above it and then drill a hole or cut a notch in the nogging to allow the cable to pass it. If the latter is found, the bob may fall between the insulation and the plasterboard cladding; alternatively, it may be possible to push the cable down between the two; only a trial will tell.

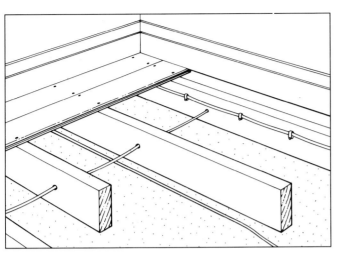

RUNNING CABLE ACROSS FLOORS AND CEILINGS

The void underneath suspended timber floors is the perfect place to hide cable runs, and if the house has a loft, access to it for carrying out modifications or extensions to the lighting circuits serving the floor below could not be easier.

The time to run cables beneath suspended timber floors is when the room is being redecorated and fitted floorcoverings can be removed to allow the floorboards to be lifted for access.

If you want to run cable parallel to the joist direction and there is a ceiling beneath the floor, all you have to do is lift a floorboard at each side of the room and push the cable across the ceiling. You will only encounter difficulties if there is strutting between the joists or if the ceiling is of lath-and-plaster. In the first situation, you will simply have to lift every third or fourth board so you can feed the cable between the struts. In the second, the cable may snag on the ridged plaster; either use a length of stiff wire as a pull-through, or tape the cable to a makeshift 'toboggan' such as a plastic picnic plate or a Frisbee, which will skate over the ridges as you push the cable across the ceiling surface.

Where there is a void beneath a timber ground floor sufficiently deep to use as a crawl space, lift enough boards to allow you to gain access and then clip the cable to the sides of the joist at the spac-

ings given on page 59. Alternatively, lift boards at intervals to clip the cable to the joist sides. You can simply let the cable rest loose on the concrete beneath the floor if this is present.

To run cable at right angles to the joist direction, you need to lift only one board. Drill a hole in the centre of each joist, at least 50mm (2in) from the top, with a joist brace or electric drill and feed the cable through. Wherever possible drill the holes in a zone between ¼ and ⅔ of the way across the joist span; avoid drilling into the ends or the middle section of a joist.

You will need to use a floorboard saw, jig saw or circular saw to free tongued-and-

grooved boards and chipboard floors. Punch the fixing nails through chipboard panels before trying to lift them: they may snap if you try to prise them up. Replace lifted boards with screws rather than nails, and label them 'CABLES UNDER' for future reference.

In lofts, clip the cables to the joist sides above any loft insulation; if the cable has to cross them at any point route it so it cannot be trodden on by anyone in the loft.

Lastly, if you have solid floors, do not attempt to cut cable channels in them; you may damage the damp-proof membrane beneath the floor surface. Instead run cables round the walls using mini-trunking or skirting trunking.

THINGS YOU NEED

- Floorboard saw, jig saw or circular saw
- Bolster chisel
- Claw hammer
- Joist brace or electric drill plus wood bit
- Cable clips
- Woodscrews
- Screwdriver
- Pencil

CONNECTING CABLE TO WIRING ACCESSORIES

It is essential that wiring accessories are connected properly. Bad connections can lead to short circuits and overheating, and could even start an electrical fire or give a shock to anyone using the accessory.

THINGS YOU NEED

- Wiring accessory
- Side cutters
- Wire strippers
- Green/yellow PVC earth sleeving
- Red PVC insulating tape
- Handyman's knife or scissors
- Terminal screwdriver

It goes without saying that each cable core must be connected to the correct terminal on the accessory faceplate, yet many installations are found to have reversed polarity – live and neutral connections transposed – or, worst of all, to have the live and earth connections reversed. Make sure that you can clearly identify each terminal on the accessory concerned, and that you connect live (red) cores to terminals marked L or coloured red, neutral (black) cores to those marked N or coloured black, and the earth cores to those marked E, ⏚ or coloured green.

Next, always leave a generous amount of cable at each connection point; it is easy to cut off any excess, but if the cable is too short, making the connections to the terminals can be very difficult.

Do not remove too much core insulation; you need enough bare conductor to make a good connection within the terminals, but no more. None should be visible once the core is connected. Always cover the bare earth core with green/yellow PVC sleeving before connecting it to its terminal. Twist like cores together if you are connecting them to the same terminal.

When you have made the connections, fold the excess cable carefully back into the mounting box without kinking it. If the cable run to the box is free (not anchored in a chase, for example), push the excess cable into the space.

Stripping Cable

1 Lay the cable on a flat surface and slit the sheathing along its centre line with a sharp knife. Peel it away from the cores and cut it off neatly.

2 Cut the cores to the length required with side cutters, then remove about 10mm (3/8in) of the insulation from each core using wire strippers set to match the core size. If you cut into the core as you do this, cut it off and start again. Remember to cover the bare earth core with green/yellow PVC sleeving before connecting it to its terminal.

1 At socket outlets and other power circuit accessories, connect the cores to their labelled terminals. If using a metal mounting box, cut a short length of cable, pull out its earth core, cover this in green/yellow PVC sleeving and use it to link the faceplate earth terminal with the brass earth terminal in the box.

2 Accessories fitted to plastic mounting boxes do not need an earth connection between faceplate and box. This socket outlet has separate terminals (for two ring circuit cables cores) which are easy to get at and are helpfully colour-coded.

3 At light switches, the cable earth core goes to a brass terminal in the mounting box unless the faceplate is metallic; in that case the core goes to the earth terminal on the switch faceplate, and if the mounting box is metal an earth link is fitted between the faceplate and the box as in Step 1. How the live and neutral (switch return) cores are connected to the faceplate terminals depends on the type of switch and on whether it is being wired up for one- or two-way switching; see pages 75 and 84 for more details of switching arrangements.

4 At ceiling roses, feed the cable through the rose baseplate before fixing it to the ceiling. Which terminals you use depends on whether you are using loop-in or junction-box wiring (the latter is shown here; see page 77 for more details). Then push the cable back into the ceiling void and screw the baseplate to the ceiling, ready for the pendant flex to be attached and the rose cover to be fitted.

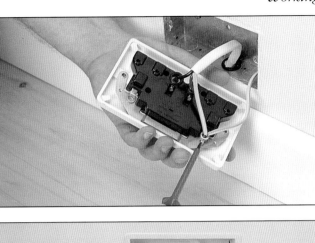

Step 1

Step 2

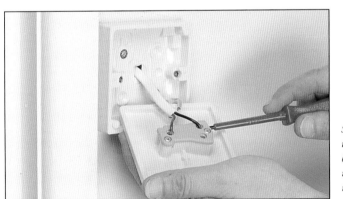

*Step 3;
remember to tag the
black switch core with
red tape before re-
mounting the switch*

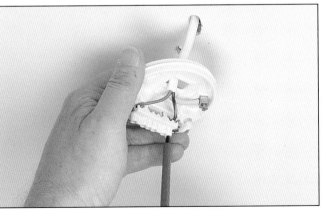

Step 4

EARTHING AND CROSS-BONDING

Earthing and cross-bonding are essential in any wiring system. They provide a safe path through which electricity can flow harmlessly to earth if a fault develops.

THINGS YOU NEED

- Single-core PVC-sheathed earth cable in size to suit situation
- Earth clamps for pipework
- Earth terminals for sinks and baths
- Side cutters
- Wire strippers
- Screwdriver

As explained on page 14 every part of the system must be linked to earth, with a continuous protective conductor running round each circuit and back to the main earthing terminal at the consumer unit or fusebox. This is then linked via a main earthing conductor either to the sheath or neutral core of the incoming supply cable, or to an earth electrode.

Old wiring systems may not have an earth conductor on the lighting circuit, and if you find this to be the case one should be added at the earliest possible opportunity – certainly before any other work is done on the circuit. This involves running a continuous 1mm² single-core earth cable from the main earthing terminal to every lighting point and switch on the circuit. If any existing lighting points have no earth terminals they must be replaced by fittings that do.

MAIN EQUIPOTENTIAL BONDING CONDUCTORS

These are single-core earth cables with green/yellow sheathing that link the metalwork of services such as gas and water supplies to the system's main earthing terminal. They must have a cross-sectional area of at least 10mm² and be attached to the gas and water pipes by a special earth clamp made to BS5951. On the water supply pipe the connection should be made as close as possible to the pipe's point of entry to the property – immediately after the stoptap if the supply pipe is plastic, before it if the supply pipe is made of metal, and in both cases before any junctions. The connection to the gas pipe should be made within 600mm (2ft) of the meter, on the consumer's side of it. In older houses there are other metal pipes which go out of the property into the ground, such as lead waste pipes and cast iron soil pipes. These must also be linked to the system's main earthing terminal with 10mm² cable.

SUPPLEMENTARY BONDING CONDUCTORS

Supplementary bonding or cross-bonding conductors are also single-core sheathed cables. They link to earth any exposed metalwork that is not part of the wiring system, but which could accidentally come into contact with it and so could become live itself. This includes things such as metal sinks, central heating pipework and also all exposed metalwork in bathrooms. Earth clamps are used for cross-bonding connections to pipework; metal sinks and baths usually have a special earth tag attached to them to take the connection (if yours do not, simply drill a hole and fit a small nut and bolt to take the connection). The cross bonding conductor is then run back to the system's main earthing terminal.

The correct size for a supplementary bonding conductor is tricky to work out. The minimum size is 2.5mm² if it is sheathed or otherwise protected, but most electricians use 4mm² cable throughout the system.

In a properly earthed system the gas and water supply pipes and any other exposed metalwork are bonded to earth. Check that your system has all these main and supplementary bonding conductors; install any that are missing

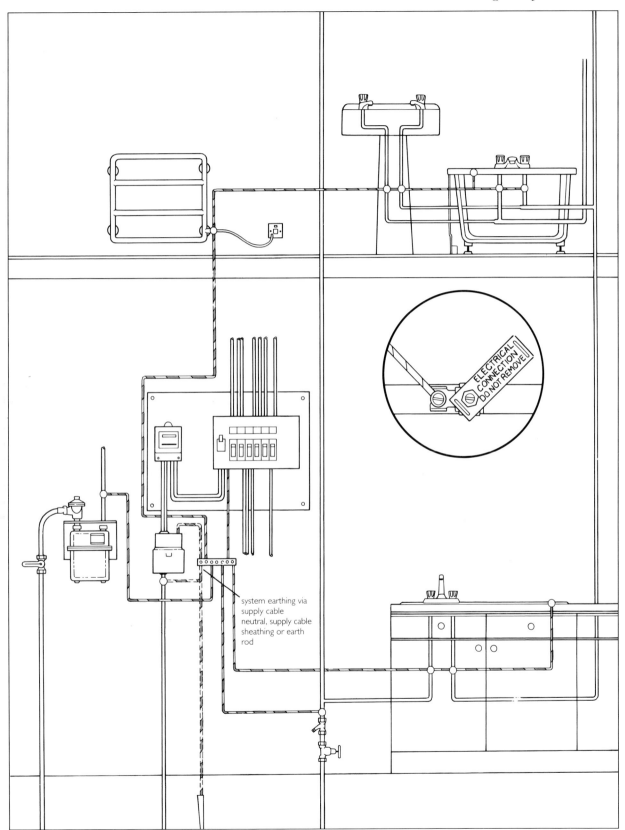

system earthing via
supply cable
neutral, supply cable
sheathing or earth
rod

ELECTRICAL
CONNECTION
DO NOT REMOVE

MOUNTING WIRING ACCESSORIES

Once a circuit cable reaches its destination, it is connected to the wiring accessory it serves. All these accessories – light switches, socket outlets, ceiling roses and so on – must be securely fixed to the wall or ceiling surface. In addition, the connection between the cable and the accessory must be made within a suitable enclosure or mounting box complying with the relevant British Standard, so that the wiring is inaccessible when the accessory is fixed to the box.

This enclosure may be surface- or flush-mounted, and is made of rigid plastic, galvanized steel or aluminium. Special weakened areas called knockouts are formed in the back and/or sides of the box, and are designed to be removed during installation to allow the circuit cable to enter the box from whichever direction is appropriate. The knockouts in plastic boxes are thinned areas which can be broken out with a screwdriver. In metal boxes the knockout is a pre-punched circular disc which is simply pushed out and broken off, and the resulting hole must be lined with a protective rubber washer known as a grommet to prevent the cable sheath from chafing on the metal edges.

Metal boxes contain an earth terminal which must be linked to the earth terminal on the wiring accessory by a short sleeved earth wire. Plastic boxes do not need an earth terminal, but one may be fitted in boxes for plastic plate-switches with no earth terminal, to provide a termination point for the switch cable earth core; switches with metallic faceplates have an earth terminal to take the earth core.

SURFACE-MOUNTING ACCESSORIES

Surface-mounting is the quickest and easiest way of installing wiring accessories. It is particularly popular for wiring in garages, workshops and the like, where neatness of appearance is less important than around the house, often in conjunction with tough metal-clad wiring accessories.

The mounting box is simply screwed to the wall surface at the desired position, and the circuit cable is run into it through one of the knockouts from the back (concealed wiring) or from the side (surface wiring or wiring run in mini-trunking). The one drawback with surface-mounted accessories – especially the deeper types such as cooker control units and shaver supply units – is that they project noticeably from the wall, and so are more prone to accidental damaged than flush-mounted accessories.

However, you may be prepared to put up with this disadvantage as an alternative to the task of chopping holes in your walls for recessed boxes, or as a short-term measure for new wiring work until you next redecorate the room and can flush-mount both the wiring and the accesories.

Decide where you want the new box to be mounted, and hold it against the wall so you can mark the positions of the fixing screws through the holes in the back of the box.

One of these may be a vertical slot to allow the box to be levelled as it is fitted, but it is a good idea to use a small spirit level as an aid to accuracy at this stage. If you are mounting the box on a stud partition wall, try to fix it to one of the studs if possible; locate these with an electronic stud finder. Cavity fixing devices do not provide a strong enough fixing in plasterboard.

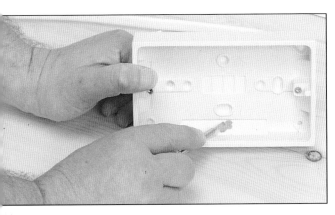

2 Work out how the cable will enter the box, and press out the appropriate knockout using the tip of a screwdriver. If you are using mini-trunking, the special mounting boxes usually have a small removable rectangular panel in the side of the box; slide this out and clip in an adaptor to link the box to the mini-trunking.

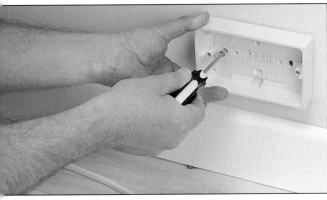

3 Drill the fixing holes, insert wallplugs into a masonry wall and screw the box into place. For flush wiring, feed the cable in through the rear knockout first. Use your spirit level again to ensure that the box is level before tightening the screws fully. For surface wiring or wiring with mini-trunking, feed the cable in after you have secured the mounting box.

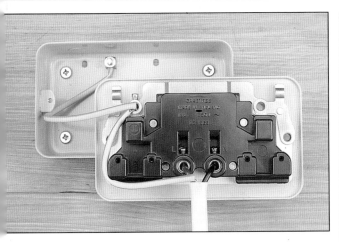

Metal-clad accessories and steel mounting boxes are ideal for surface mounting in garages and workshops

FLUSH-MOUNTING ACCESSORIES IN SOLID WALLS

The neatest way of mounting your wiring accessories is over a box that is recessed into the wall so that the faceplate fits flush with the wall surface. The cable, run in a chase to the box position, enters the box via a knockout in the side or rear face as appropriate.

THINGS YOU NEED

- Mounting box
- Rubber grommets
- Pencil
- Straightedge
- Small spirit level
- Electric drill plus masonry bits
- Drilling guide (optional)
- Sharp brick bolster or cold chisel
- Club hammer
- Woodscrews and plastic wallplugs

The mounting boxes used for flush-mounting wiring accessories in masonry walls are usually made from galvanized steel (to BS4662 for plateswitch boxes, to BS4662 or BS5733 for other boxes), although aluminium is sometimes used for the extra-deep boxes (to BS4177) required by cooker control units. They have pre-drilled fixing holes, knockouts in the back and in all four sides, a brass earth terminal and two fixing lugs to which the accessory faceplate is screwed; one of these may be adjustable to allow the faceplate to be fixed squarely even if the box is slightly out of alignment.

Top to bottom:
Steps 2, 3, 4 and 5

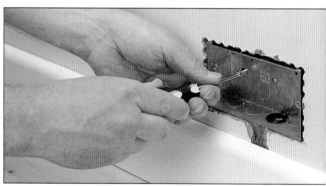

Select the correct box for the accessory you are installing – 16mm (⅝in) deep for plateswitches, 25 or 35mm (1 or 1½in) for most power circuit accessories. Hold the box against the wall at the selected fixing position, use the spirit level to check that it is level and draw round it in pencil.

2 You can make it easier to chop out the recess accurately if you drill a honeycomb of holes in the wall first. You can do this freehand, or use a proprietary drilling guide. Hold it over the box outline, mark the positions of its fixing holes, drill these and screw the guide to the wall. Then mark the overall drilling depth needed – guide plus box depth – on the drill bit, and drill out each hole in turn; the guide can be used for both single and double mounting boxes. When you have finished drilling, remove the guide.

3 Chop out the plaster and masonry to the required depth using the brick bolster and club hammer. Work from the edges towards the centre of the recess to minimise damage to the surrounding plaster and masonry. If you are cutting deep recesses in walls one brick/block thick, do not be too heavy-handed or you may knock a hole right through the wall.

4 Test the fit of the mounting box, then mark the positions of its fixing holes. Drill them out, fit wallplugs and screw the box to the masonry after removing a knockout at the cable entry point and fitting a grommet.

5 Feed the cable into the box, ready for connection to the accessory faceplate.

There is now a range of plastic mounting boxes specially designed for flush-mounting wiring accessories in stud walls. They have a narrow flange that fits against the face of the plasterboard (and also disguises any minor inaccuracies in the edges of the cutout), and are fitted with spring-loaded or rotating clips that grip the rear face of the board securely when the box is pushed into place.

Note that these boxes may not work on old lath-and-plaster walls; it depends on the depth of plaster used. If the plaster is too thick, either surface-mount the box over a stud or flush-mount it in a notch cut in one of the studs.

Check that the chosen fixing position does not coincide with a stud by using a stud finder, or by tapping the wall surface to check that it sounds hollow. Then hold the box against the wall, using a spirit level to help position it straight, and draw round its outline in pencil.

2 Drill a hole through the plasterboard just inside each corner of the outline, big enough to admit the tip of the padsaw. Insert it in one of the holes and cut carefully along the marked lines from hole to hole. Remove the offcut.

3 Feed the cable up or down to the cut-out and pass it through a knockout in the mounting box. Then press the box into place in the cutout so its flange fits flush with the wall surface. If it has spring-loaded fixing lugs, check that these have engaged properly. If the lugs are the rotating type, turn them so that they lock securely against the rear face of the plasterboard.

FLUSH-MOUNTING ACCESSORIES IN STUD WALLS

Many homes have internal walls built as timber-framed stud partitions clad with plasterboard. It is relatively easy to conceal cable runs within these walls, and flush-mounting wiring accessories has now been made equally simple with the introduction of the flanged cavity fixing box.

THINGS YOU NEED
- Mounting box
- Pencil
- Small spirit level
- Straightedge
- Electric drill plus masonry bit
- Padsaw
- Screwdriver

Top to bottom:
Steps 1, 2 and 3

MOUNTING ACCESSORIES ON CEILINGS

Ceiling roses, cord-operated ceiling switches and some types of light fitting must, like wall-mounted accessories, be securely fixed in place and must also have their electrical connections contained within a suitable enclosure.

THINGS YOU NEED

- Ceiling rose *or*
- Mounting box to suit accessory *or*
- Round conduit box to BS4568
- Woodscrews
- Screwdriver
- Pencil
- Padsaw
- Scrap wood for batten and blocks
- Tenon saw or jig saw
- Electric drill plus drill bits

In most homes the ceilings consist of plasterboard nailed to the underside of the ceiling joists, while older properties may have lath-and-plaster ceilings. Neither is strong enough to support light fittings or pull-cord switches: these must either be screwed to the underside of a joist or to a supporting batten fitted between the joists above the ceiling surface. Except in rooms with a loft above, this will mean gaining access to the ceiling void by lifting floorboards in the room above.

As far as providing a suitable enclosure for the wiring connections is concerned, there are two options. The first is to use only surface-mounted accessories, fitted over a mounting box or block that is screwed to the ceiling surface in much the same way as a surface-mounted accessory on a wall. Modern ceiling roses are designed as surface-mounted accessories. The second, used to flush-mount ceiling switches and luminaire support couplers (see pages 78–9) and to install certain types of light fitting, is to recess the mounting box in the ceiling by fixing it upside-down to the underside of the support batten so its lower lip is flush with the ceiling surface in the room below. The mounting box can be of metal or plastic.

Left to right:
Steps 2, 3 and 4

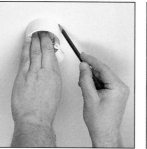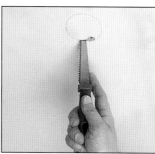

I Attach the baseplate of a ceiling rose to the underside of a joist wherever possible; the knockouts are offset and so can be positioned in such a way that the cable can run down the side of the joist and into the rose. Use the same technique to fit a surface-mounted box for a ceiling switch.

2 To flush-mount an accessory or fitting, select a fixing position mid-way between two joists. Check that there are no cables lying on the ceiling above. Then hold the mounting box or conduit box against the ceiling and draw round it in pencil.

3 Drill a hole through the ceiling within the outline large enough to admit the tip of a padsaw blade. Use the saw to cut round the outline, and remove the cut-out.

4 Gain access to the space above the ceiling, and cut a batten wide enough to fit between the joists. Screw or nail a block to each end of the batten, then fix these blocks to the joist sides to secure the batten. Note that it should be fixed just above the level of the ceiling, so that when the mounting box or conduit box is screwed to its underside, the lip of the box will be flush with the ceiling surface in the

room below. Use a conduit box with a rear rather than a side entry hole on lath-and-plaster ceilings.

5 Screw the mounting box or conduit box securely to the underside of the batten from below, and feed in the circuit cable ready for connection to the accessory or light fitting. Write 'LIGHT UNDER' on the floorboards after replacing them.

WIRING JOBS

WORK ON LIGHTING CIRCUITS
WORK ON SOCKET OUTLET CIRCUITS
WORK ON OTHER CIRCUITS
INSTALLING NEW CIRCUITS
WIRING OUT OF DOORS
MAINTENANCE AND REPAIRS

WORK ON LIGHTING CIRCUITS

Before you start doing any work on your lighting circuits, take the time to survey them so you know how they work and, more importantly, that they are safe to work on. Start by establishing what circuits you have and what areas each one serves, by the simple expedient of removing each circuit fuse or switching off its circuit breaker, and then operating the switches. Keep a written record of your findings. Make certain that none of the circuits is overloaded: there should be a maximum of eight light fittings on each.

Next, work out how each lighting point is wired up and controlled, especially if there is

any two-way switching or if individual switches control more than one light. With the power off, unscrew ceiling roses and ceiling-mounted light fittings to reveal the wiring inside. Check first of all that the circuit has an earth conductor; old wiring may not have one, and you will have to add one all the way round the circuit if this is the case.

A single circuit cable at a ceiling rose or other light fitting indicates that it is fed from a junction box or as a spur, or that it is the last light on the circuit, while two or more circuit cables show that it is wired using the loop-in system (see pages 38–40 for more details).

At switches, note how many gangs there are, and whether the switch is a one-way type (two terminals per gang) or two-way (three terminals). Check whether two-core or single-core cable has been used for the switch drop, and identify two-way or intermediate wiring arrangements.

Again record your findings for future reference, ideally on a simplified floor plan of the house. Mark the position of individual lighting points and their switches on it by numbers, and then add corresponding marginal notes about each one. You will find them a great help for all your future wiring work.

REPLACING SWITCHES

The simplest of all wiring jobs is the replacement of an existing plateswitch, either for cosmetic reasons or because it is damaged or faulty. You can fit another plateswitch, perhaps in a different style or material, or take the opportunity to install a dimmer switch so you can control the light's brightness.

Right: A one-way switch (top) has two terminals; a two-way switch (bottom) has three, but only two of the three are used if such a switch is wired for one-way switching. Before fitting the switch to its mounting box flag the black core with red PVC tape to show that it is live

The first thing to determine is what type of switch it is, if you have not carried out the survey described above, so you can buy the right replacement.

REPLACING A PLATESWITCH

I Turn off the power at the consumer unit and make sure the circuit is dead. Undo the screws securing the faceplate; with some 'decor' switches, these may be hidden behind a clip-on fascia which you will have to prise off first. If the switch is flush-mounted, run a sharp knife round the faceplate to cut through any paint or wallcovering stuck to its sides. Then ease it away from the wall to see whether it is a one-way or two-way switch. It will be obvious how many gangs it has.

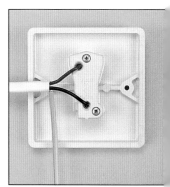

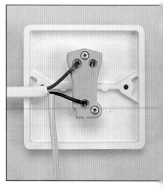

2 Next, draw a simple sketch showing which core is connected to which terminal, especially if any contain more than one core, and what earthing arrangements (if any) are present.

3 Disconnect the cable cores from the faceplate. If the incoming cable earth core goes to a terminal on the mounting box, leave it there (but if it is bare, disconnect it, slip on a piece of earth sleeving and reconnect it). Discard the old switch.

4 Reconnect the cores to the terminals on the new switch faceplate, reproducing the connections recorded on your sketch. If two-core-and-earth cable has been used, identify the black core(s) as live with a piece of red PVC insulating tape. Do the same for the blue and yellow cores on three-core-and-earth cable.

5 If you are fitting a metal faceplate on a metal mounting box, link its earth terminal to the box terminal with a short length of earth core, taken from an offcut of 1mm² cable and covered with a piece of PVC sleeving.

6 Carefully fold the cable(s) back into the mounting box and screw the faceplate to it. If you have difficulty getting the new screws to engage in the fixing lugs, this may be because they are metric and have a different thread pattern to that in the lugs; try using the screws from the old switch instead. Then restore the power and test the switch.

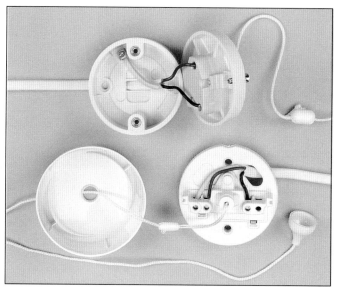

FITTING A DIMMER SWITCH

Dimmer switches come with detailed wiring instructions; simply follow these when fitting one in place of an existing plateswitch.

REPLACING A CEILING SWITCH

If you are replacing a cord-operated ceiling switch, undo its fixing screws and ease it away from its mounting box. Draw a sketch of the connections as before, then disconnect the cores, reconnect them to the replacement switch and reattach it to its mounting box.

Top: Two types of one-way ceiling-mounted switch

Left: Dimmer switches are available in one-gang and two-gang versions. Remember to flag the black switch core(s) with red PVC tape

REPLACING A PENDANT FLEX

Fitting a new length of flex between a ceiling rose and its lampholder involves nothing more complicated than unscrewing the cover of each component, disconnecting the old flex and fitting a new length in its place. It's easier than wiring a plug

THINGS YOU NEED

- Replacement heat-resistant flex
- Terminal screwdriver
- Side cutters
- Handyman's knife
- Wire strippers
- Replacement lampholder (optional)

Below: Plastic lamp holders are wired with two-core flex. Each core is hooked over a small anchor as shown

1 Turn off the power at the consumer unit and check that the circuit is dead; turning off just the light switch will not isolate the rose safely. Then unscrew the rose cover. If it is stuck fast with paint, run a knife round its base to free it.

2 Let the rose cover slide down the flex, and make a note of which flex core goes to which terminal. Undo the screws securing the flex cores and unhook the cores from their anchors. Lift the lampholder and flex down. Discard both if you are fitting a new lampholder.

The flex linking a ceiling rose to its lampholder may need replacing for several reasons. It may be the old twisted-core type with clear insulation, which looks old-fashioned and is no longer allowed. It may have become discoloured with age, or by cigarette smoke. It may be too short, or it may have stretched so that the flex cores are visible, especially if these were not hooked over their anchors or too heavy a lampshade has been used. Worst of all, it may have become worn by the light swaying to the point where the insulation breaks down and a short circuit occurs.

Make sure that you use the correct replacement flex –

3 Cut the replacement flex to the required length, and strip about 38mm (1½in) of the sheath from one end. Prepare the cores and connect them to the two lampholder terminals. With metal lampholders, connect the flex earth core to the earth terminal on the lampholder. Then hook each core round its anchor and screw on the lampholder cover. Make sure the skirt is hand-tight.

4 Strip about 50mm (2in) of sheath from the other end of the flex and prepare the cores as before. Thread the

round two-core flex for plastic lampholders, three-core flex for metal ones which need earthing. Choose 0.5mm² flex unless you are installing a heavy lampshade or fitting; this size will safely carry only up to 2kg (4lb). The 0.75mm² size can support up to 3kg (6lb), and 1mm² flex up to 5kg (11lb).

If the existing lampholder is cracked, discoloured, brittle due to heat from the lamp (which can also 'lock on' the skirt that holds the lampshade), replace it at the same time as fitting new flex. Check that it is a heat-resistant type, made to category T2 of BS5042, and use heat-resistant flex.

flex through the rose cover, checking that it faces the right way. (Forgetting to do this will be very annoying when you complete step 5.)

5 Connect the flex cores to the rose terminals and loop each one over its anchor. With three-core flex, take the earth core to the rose's earth terminal. Then fit the rose cover. If the cores are too long and can be seen when the cover is on, disconnect them, shorten them slightly and reconnect them. Restore the power and test the light.

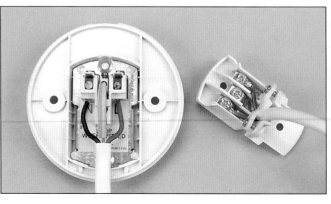

Right: At a luminaire support coupler (see over) the flex is connected into a plug-type fitting with a cord grip. Flex connections to a ceiling rose are shown opposite

Modern plastic ceiling roses are designed for surface-mounting, either to the undersides of ceiling joists or to a support batten fixed between adjacent joists. The circuit cables enter through the base of the rose and are connected to a bank of terminals, arranged in a line and angled for ease of connection. These are divided into three groups, with barriers between each group; the centre group is for loop-in wiring, and is not used if the rose is fed from a junction box.

An alternative is the pillar-terminal type. This more closely resembles a junction box internally, with either three separate pillar terminals plus an earth terminal for loop-in wiring, or two plus an earth if it is designed for junction-box wiring.

Removing an old ceiling rose may reveal ceiling damage round the rose site; this can be concealed by fitting a circular plastic halo round the base of the rose.

It is still possible to buy semi-recessed porcelain roses, which may be preferred in restored period properties. They are designed for installation on a standard round conduit box recessed into the ceiling surface.

FITTING A NEW CEILING ROSE

If your home has old-fashioned, discoloured or paint-encrusted ceiling roses, you may want to replace them, perhaps with brass fittings as an alternative to the ubiquitous white plastic type.

1 Turn off the power at the consumer unit and check that the circuit is dead. Unscrew the old rose cover and make a sketch of where the circuit cable and pendant flex cores go. Disconnect them, labelling them as you go, unscrew the rose base from the ceiling and discard it.

2 Remove a knockout or two as necessary from the base of the new rose, feed in the cables and screw the rose to the ceiling.

3 Reconnect the circuit cables to the new rose, reproducing the connections as per your sketch. Two typical arrangements are shown here. Cover any bare earth cores you find with a piece of green/yellow PVC sleeving.

4 Either disconnect the flex and its lampholder from the old rose and reconnect them to the new one, or fit a new flex and lampholder (see opposite). Screw on the rose cover and restore the power.

Left: At the last loop-in rose on a circuit (top) there is one circuit cable and a cable to the switch (identified by a red PVC flag on its black core). At other roses on a loop-in circuit (bottom) there are two circuit cables along with the switch cable

THINGS YOU NEED
- Replacement ceiling rose
- Screwdriver
- Terminal screwdriver

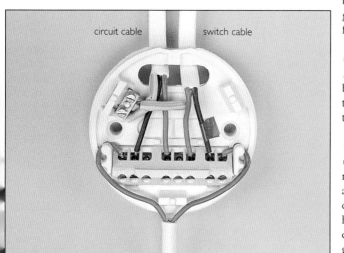

circuit cable   switch cable

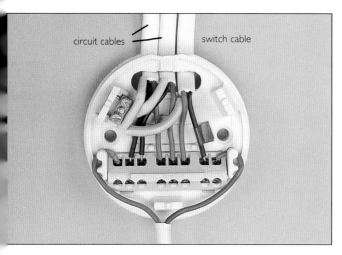

circuit cables   switch cable

FITTING A LUMINAIRE SUPPORT COUPLER

You might like to take advantage of modern lighting technology by installing plug-in ceiling lights in the place of ceiling roses. These are known as luminaire support couplers (LSCs).

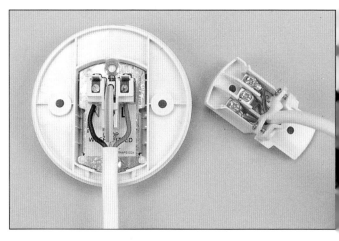

THINGS YOU NEED

- Luminaire support coupler
- Screwdriver
- Terminal screwdriver
- Mounting box for LSC
- General d-i-y tools if you are flush-mounting the LSC

Having pendant lampholders and other suspended fittings permanently connected to the mains is a nuisance, making it difficult to take them down when decorating or for cleaning or replacement. By replacing existing ceiling roses with plug-in types, fittings can be connected and disconnected as easily as plugging in an electrical appliance

The socket part of the rose is wired up in the same way as an ordinary rose, using either loop-in or junction-box wiring, and is fitted over a round conduit box for flush installations – best avoided unless you can gain access easily to the ceiling void to fit the box – or on a round mounting box for surface-mounting. The pendant flex is connected to a special plug which engages in the socket; three different types are shown. A screw-on cover conceals everything (and in one design, locks the plug in place too), resulting in a neat installation that looks very similar to a traditional ceiling rose. One manufacturer also offers an adaptor socket which can be mounted directly over an existing ceiling rose.

LSCs will generally support heavier weights than an ordinary ceiling rose – 5kg (11lbs), or even more.

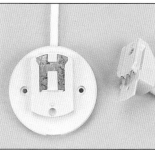

I Turn off the power at the consumer unit and check that the circuit is dead. Unscrew the old rose cover and make a sketch of where the circuit cables cores go. Disconnect them, labelling them as you go, unscrew the rose base and discard it.

2 For surface-mounting, remove a knockout or two as necessary in the base of the mounting box, feed in the cables and screw the box to the ceiling.

3 Reconnect the circuit cables to the LSC socket as per your sketch. Cover any bare earth cores you find with green/yellow PVC sleeving.

4 Thread the rose cover on to the flex, making sure it faces the right way. Then connect the pendant flex to the plug, insert the plug in the socket and screw on the cover.

Above and left: LSCs made to BS7001 have interchangeable slide-in pins, so lights can be swapped around

Below: earlier types of plug-in rose have a round-pin plug or a clip-on connector

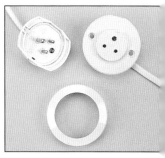

There is a huge range of decorative light fittings on the market, but they are connected to the mains in one of just two ways, depending on how the original ceiling rose they are replacing was wired up. They are designed to be screwed to the ceiling surface – either to a joist or to a supporting batten – and the circuit connections are made either to a terminal block within the body of the light fitting or within a suitable enclosure such as a round conduit box recessed into the ceiling, which is then covered by the fitting's baseplate.

Fittings with a terminal block can usually be fed by only a single cable, so you have to use junction-box wiring. The one exception is the **battenholder**, a utility fitting designed for use in garages, lofts, cupboards and so on; some of these can be be connected to loop-in wiring too. Other fittings have flex tails already attached, ready for connection to the mains cable via strip connectors, and can be connected using either loop-in or junction-box wiring.

Some light fittings are designed so that the baseplate can be screwed directly to the fixing lugs of the conduit box (these are at a standard 51mm spacing) using the same 3.5mm (M3.5) machine screws used to attach other wiring accessories to mounting boxes. Others have larger baseplates and are screwed to the ceiling over the box, so will generally need a supporting batten.

I Turn the power off at the consumer unit and check that the circuit is dead. Unscrew the ceiling rose cover. Label the cable cores as you disconnect them from their terminals. Unscrew the ceiling rose and discard it.

2 Hold the conduit box against the ceiling and draw round it. Drill a hole just inside the outline and use a padsaw to cut round it.

3 Gain access to the ceiling void above to check on the rose's position. If it is screwed to a joist, you will have to fit a support batten between the joists in order to fit the recessed box. If the rose is screwed to a batten, this will have to be repositioned so the lip of the box finishes flush with the ceiling surface.

4 Screw the box to the batten from below, then feed the cable(s) into it. Connect the cores to strip connectors according to your labelling. Two examples are shown.

5 Connect the flex cores from the light fitting to the strip connectors. Then push the connector strip into the conduit box and screw the light fitting in place, to the box or the ceiling.

6 Alternatively, connect the circuit cable(s) direct to the terminal block of fittings such as fluorescent lights and battenholders, then fix them to the ceiling.

INSTALLING A NEW LIGHT FITTING

Pendant lampholders do not allow much scope when it comes to making the most of your lighting's looks. For the sake of both appearance and performance you will probably want to install some more decorative – and functional – fittings.

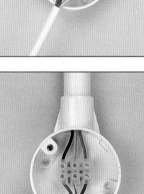

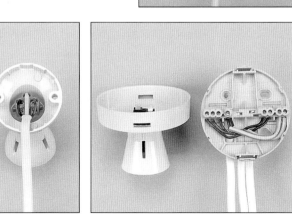

Above: if conduit boxes are required, use strip connectors to link the light fitting to the mains. The connections shown are for the end of a loop-in circuit (top) and junction-box wiring (bottom)

Left: Battenholders are available for junction-box wiring (left) or for loop-in wiring (right). The battenholder on the right is plugged into its base after the wiring

EXTENDING LIGHTING CIRCUITS

If you want extra lights in new positions, you can extend an existing lighting circuit to supply them. This is easy to do so long as you have access to the existing circuit cables and you will not overload the circuit as a result of adding the extra lights.

THINGS YOU NEED

- 1mm² two-core-and-earth cable
- Green/yellow PVC earth sleeving
- Junction boxes to suit wiring option
- New ceiling rose or light fitting
- New plateswitch and mounting box
- Handyman's knife
- Side cutters
- Wire strippers
- Screwdriver
- Terminal screwdriver
- Red PVC insulating tape
- General d-i-y tools

1: Spur to supply a new light connected to the circuit cables at an existing junction box
2: Spur to supply a new light connected to the circuit cables at an existing ceiling rose
3: Connections for a new light at a four-terminal junction box cut into the circuit cable
4: Spur from a three-terminal junction box cut into the circuit cable. The new light will be looped in

Option 1:
TAKING A SPUR

You can connect a spur cable into either an existing four-terminal junction box or a loop-in ceiling rose, so long as there is enough space within it to do so. A round junction box will usually accept up to six cables, so assuming that four are already present (two circuit cables plus cables to a light and its switch) you can add two spurs. A loop-in rose can accept up to four cables, so can generally supply just one spur.

The spur runs from the main cable terminals at the original junction box or rose as shown in illustrations 1 and 2 below either to a new four-terminal junction box and on to the new rose/light, or directly to a loop-in rose/light. The switch cable is connected to the junction box in the first case, to the rose/light in the second.

Option 2:
ADDING A JUNCTION BOX

The second method is to cut into the main circuit cable at a convenient point and to wire in a new junction box at that point. If a four-terminal box is used, cables are then run from the box to the new light and switch positions (illustration 3).

If a three-terminal box is used, a spur cable is run to the new light position (illustration 4) and the switch cable is looped in there.

Which method you select will depend on a combination of convenience and making the most economical use of cable.

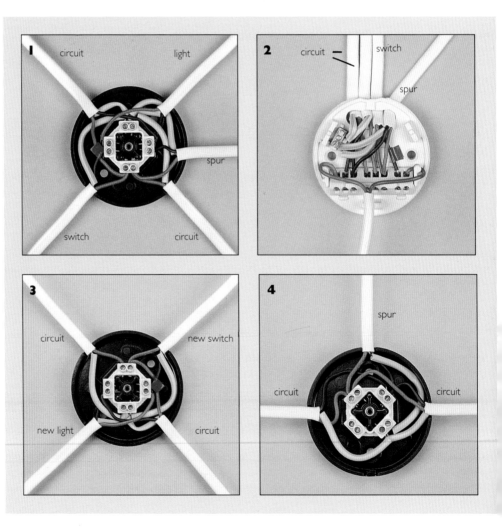

DOING THE WORK

1 Lift floorboards or survey the loft space to locate the most convenient point for your spur connection, and decide which wiring option to use. If you decide to cut a junction box into a circuit cable, trace the cable route carefully to ensure that it is not a switch cable or one running to a single light from a junction box.

2 Install the new fittings and run cables from the new light and switch positions back towards the chosen connection point – an existing junction box or loop-in rose, or a new junction box, according to the wiring option chosen. Turn the power off at the consumer unit and check that the circuit is dead. Make the connections to the circuit.

Wiring Regulations

Remember that you must first check whether adding extra light(s) will overload the circuit. Although a 5-amp lighting circuit can in theory supply up to 12 lighting points, each rated at 100 watts, in practice it is best to restrict each circuit to eight lighting points, to allow for the use of high-wattage lamps or fittings containing more than one lamp. Check how many each circuit already supplies before proceeding. If an overload will result, either check another lighting circuit for suitability if you have more than one circuit, or wire the new lights as a spur from a power circuit instead (see Option 4 in Installing wall lights, pages 81-3).

Some wall lights have integral push-button or cord-pull switches, but most are designed to be controlled by a remote switch sited elsewhere in the room. Which style and type you choose will depend on what purpose you want your new lights to serve. Once you have selected them, the main decision you have to make before you start the installation concerns how to provide a power supply and switch control for the lights; there are several options.

Assuming that potential overloading is not a problem, which option you choose will depend on convenience as much as anything. For example, in upstairs rooms it will generally be easiest to gain access to the upstairs lighting circuit where it runs in the loft, while downstairs it may be simpler to wire into a power circuit at skirting-board level, especially if the room above has fitted carpets and built-in furniture you are loath to disturb to gain access to the lighting circuit in the ceiling void.

You will be able to do much of the work with the power on, enabling you to see what you are doing and to use power tools where necessary. Turn the power off before you make the final connection to the circuit that will supply the lights.

INSTALLING WALL LIGHTS

Ceiling lights are fine for general lighting, but wall lights are much more versatile for 'mood' or task lighting, or for highlighting individual room features such as alcoves. They can also look very attractive.

Option I:

TAKING A SPUR FROM A LOOP-IN ROSE OR A JUNCTION BOX

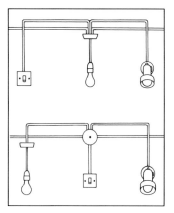

1 Turn off the power at the consumer unit and check that the circuit is dead. Wire the spur into the rose or junction box as shown below, and run it to the new light position. The light will be controlled by the existing light's switch. For independent switching connect the spur as in illustration 1 and 2 opposite. *(Continued over)*

THINGS YOU NEED

- Wall light(s)
- Round conduit box or architrave box if necessary
- 1mm² two-core-and-earth cable
- Red PVC insulating tape
- Green/yellow PVC earth sleeving
- Side cutters
- Wire strippers
- Terminal screw-driver
- Screwdriver
- Four-terminal junction box (Options 2 to 4)
- Plateswitch and box (Options 3 and 4)
- Fused connection unit (Option 4)
- 2.5mm² two-core-and-earth cable (Option 4)
- General d-i-y tools

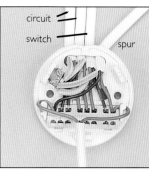

Left: Spurs connected to the switch return and neutral terminals. The wall light will come on at the same time as the existing fitting

2 The spur cable can supply several lights, so long as this will not overload the lighting circuit concerned (see box below); run it to a three-terminal junction box, and link each new light to this box by its own cable, with like cores going to like terminals as shown.

light 1
light 2
circuit

Option 2:
REPLACING A CEILING FITTING

I Turn off the power at the consumer unit and check that the circuit is dead. Remove the existing ceiling-mounted rose/light altogether, in order to use its power supply for your new wall lights. Label the cables for reference as you disconnect them.

2 Push the cable(s) up into the ceiling void and reconnect them to a four-terminal junction box (see illustration 3 on page 80), maintaining the existing circuit and switch connections. Then run a cable from the box to the new wall light(s), which will be controlled by the same switch as the existing light was. If only one cable ran to the original light fitting, you will simply be extending this cable from the new junction box (the wiring will look like that in Option 3, Step 1).

MAKING A NEW CONNECTION TO THE CIRCUIT

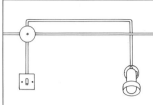

I Turn off the power at the consumer unit. Connect into an existing lighting circuit at a convenient point, using a four-terminal junction box. Make sure you connect into the main circuit cable, not a switch cable or a spur from a junction box. This arrangement allows the new lights to have their own switch. Connect the cut circuit cable ends at the box as shown below.

2 Run new cables from the box as in illustration 3 on page 80.

Wiring regulations

With Options 1 and 3 you must take care not to overload the circuit. If the circuit you choose as the power source is supplying eight existing lighting points already and you are not simply replacing one of these, you will have to select Option 4.

TAKING A SPUR FROM A POWER CIRCUIT

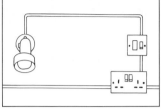

I Turn off the power at the consumer unit and check that the circuit is dead. Connect your spur cable into the circuit at any existing conveniently sited socket, so long as this is on the main circuit and is not already supplying a spur (see pages 92–3 for more details). Run the spur – which must be 2.5mm² cable – to the FEED terminals of a fused connection unit (FCU). This should contain a 5-amp fuse to provide the correct fuse protection for the new lighting sub-circuit. A switched FCU can also act as the on/off switch.

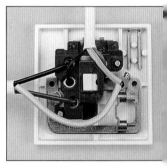

2 Use 1mm² cable for the run from the LOAD terminal of the FCU to the new lights, again using a three-terminal junction box to split the supply if you are installing more than one light (see Option 1, Step 2). If you want a conventional light switch to control the new lights, fit a four-terminal junction box first as in illustration 3 on page 80.

INSTALLING THE WALL LIGHT FITTINGS

1 Start by deciding on the precise positions of the new lights on the wall. Aim to have them at a height of about 1.8m (6ft) above the floor unless there is a reason to site them at a higher or lower level; for example, you might want to have bedside lights set just above the headboard or a reading light in the living room.

2 If you are planning to run the new cables temporarily on the surface up or down the walls, leave some slack at the point where the cable enters the floor or ceiling void, to allow it to be bedded in a recessed chase cut into the plaster at some future date.

For flush wiring (see page 61), write 'CONCEALED CABLE' on the wall alongside the chase to remind yourself or future occupiers that there is a live cable hidden in the wall.

3 If the light fitting has a hollow baseplate, simply connect the supply cable and secure the fitting's baseplate to the wall over the cable entry point with screws and wallplugs. If it needs a round conduit box or an architrave box recessed into the wall to contain the connections, chop out the recess and fit the box in place. Then feed in the supply cable, make the connections using strip connectors concealed within the box, and mount the fitting either directly to the lugs at each side of the box if they are at the standard 51mm (2in) centres; otherwise use screws and wallplugs.

4 With the new fitting connected up and mounted, run the rest of the wiring along your selected route to the point at which it will be connected to an existing circuit, then turn off the power and make the final connections. Double-check everything before restoring the power and testing your new lights.

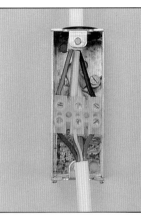

luminaire Support Couplers (lscs)

Some modern wall lights are now supplied wired to a special plug which is designed to be inserted into a matching wall socket – a combination known as a luminaire support coupler. The socket takes the place of the recessed connection box in the wall, and allows the fitting to be removed easily for replacement or during redecoration.

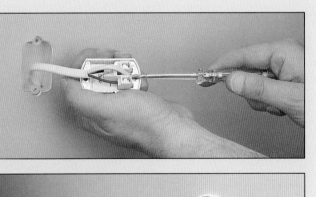

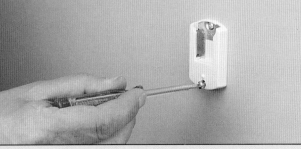

Above: Use strip connectors to link the wall light flex to the mains, either within a conduit box (the end of a loop-in circuit is shown here) or within an architrave box (junction-box wiring shown)

PROVIDING TWO-WAY SWITCHING

The simplest type of lighting control is called one-way switching, where each light is operated by its own switch. If you want to control the light from more than one switch position, you need to link the extra switches using what is known as two-way switching.

how it works

The existing switch is used as the 'master' switch. This is then linked to the second switch (and to subsequent switches if these are required) using special three-core-and-earth cable. The first switch, the second switch on a two-switch set-up and the final switch on a multiple-switch array are all two-way switches with three terminals. The other switches on multi-switch arrangements are intermediate switches, which have four terminals.

The red, yellow and blue cores of three-core-and-earth cable are so coloured for identification only; all may be live at one time or another, depending on how the switches are set. Current convention is to use the red core to link the common terminals of two-way switches, and the blue and yellow cores to act as strapping wires linking the other terminals of both two-way and intermediate switches. The blue and yellow cores should be identified as live by wrapping a strip of red PVC tape round their insulation.

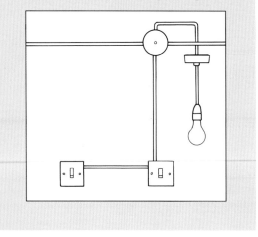

Standard two-way switching arrangement

There are several situations where being able to control a light from two (or more) switch positions is more convenient than having a single switch. The commonest is to control lights in a stairwell, where it is both more convenient and safer to be able to switch on the lights illuminating it from either upstairs or downstairs. Two-way switching is also useful in bedrooms where lights can be controlled from switches by the room door and beside the bed, in through rooms, in long corridors and for controlling outside lighting from indoors or outside.

I Turn off the power at the consumer unit and check that the circuit is dead. To provide two-way switching for a light at present under one-way control, open up the existing switch and see how many terminals it has. If it has only two it will have to be replaced, but it can be reused if it has three: two-way switches can be wired for one-way operation, but not vice versa. Buy a replacement switch if necessary, plus new two-way and intermediate switches and mounting boxes as required.

2 At the original switch position, disconnect the existing switch cable cores and reconnect them to the two terminals of the two-way switch, marked L1 and L2 if the terminal is marked C, or L2 and L3 if the one is marked L1.

Top: The master switch of a two-way switch arrangement. Centre: The second 'slave' switch. Bottom: An intermediate switch

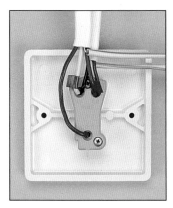

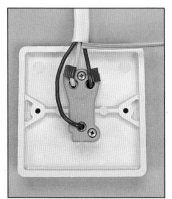

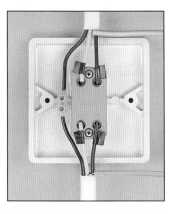

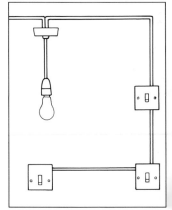

Left: Two-way switching with intermediate switch

3 If the existing switch cable is run in conduit or channelling, you may be able to feed the new three-core-and-earth cable up through it into the ceiling void above; otherwise, run the cable using the most suitable technique vertically to the floor or ceiling void. Take the cable through the void to a point vertically above or below the new switch position, and run it up or down to the new switch's mounting box. Repeat this operation to reach subsequent switches too if these are planned.

4 At the original switch position, connect the red core of the three-core-and-earth cable to the single switch terminal (C or L1, as explained in Step 2). Connect the blue core to one terminal of the pair and the yellow core to the other. Link the earth core to the earth terminal in the mounting box.

5 At the second switch position, follow the same convention – red to single, blue and yellow to the pair in the same order. Again, take the earth core to the box terminal. Add earth links between box and faceplate if either switch faceplate is metallic. Secure the faceplates to their mounting boxes.

6 If intermediate switches are being used, run cable between them as described. At the intermediate switch, connect the blue cores to the top and bottom terminals on one side, and the yellow cores to the other pair of top and bottom terminals. Link the two red cores using a strip connector. Take the earth cores to the box earth terminal.

SWITCHING IN STAIRWELLS

The ideal switching arrangement for lights in halls and on landings is the ability to turn either light on from either location. This requires the use of two two-gang two-way switches, one upstairs and one downstairs.

It is important to remember that if the two lights are on different circuits, there will be live cables at each switch position even if one lighting circuit or the other is isolated from

the mains. This is not advisable and to avoid this situation wire the landing light off the downstairs circuit.

2 The switch cable from the landing light goes to the left-hand gang of the upstairs switch; this is then linked to the left-hand gang of the downstairs switch with three-core-and-earth cable.

3 The switch cable from the hall light goes to the right-hand gang of the downstairs switch, which is then linked to the right-hand gang of the upstairs switch. The end product is two parallel two-way switching arrangements.

Below: Two-way switching for hall and landing lights. The lights are wired on the same loop-in circuit

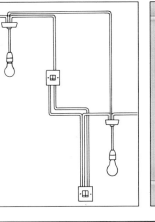

switch off first

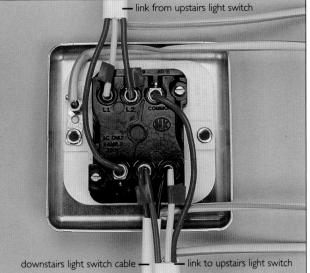

link from upstairs light switch

L1 L2 COMMON

downstairs light switch cable — link to upstairs light switch

Left: Wiring for hall (shown) and landing switches. In each case the left-hand gang is the 'slave' switch for one light. while the right-hand gang is the master switch for the other one. Above is a simpler switching arrangement found in many homes – the hall light can be switched from upstairs or downstairs, the landing light from upstairs only

FITTING RECESSED LIGHTS

While ceiling roses and many light fittings are designed to be surface-mounted, there is a wide range of fittings such as downlighters, wall-washers and swivelling spotlights which have to be recessed into the ceiling.

THINGS YOU NEED

- Light fitting(s)
- Junction box(es)
- Plateswitch and mounting box
- 1mm² two-core-and-earth cable
- Green/yellow PVC earth sleeving
- Red PVC insulating tape
- Compasses
- Padsaw
- Heat-resistant pad
- Heat-resistant core sleeving
- Electrical tools
- General d-i-y tools

Top: Basic circuit diagram
Centre: Heat-resistant pad fitted above light in ceiling void
Bottom: Protective box for light installed in loft

Right: Standard junction box wiring for a light fitting

The main advantage of this type of fitting is that they are unobtrusive and so will blend in with any type of decor. Downlighters are especially suitable for providing glare-free lighting in halls and stairwells; directional types add extra versatility; and dimmer control will let you alter the light level too.

I Lift a floorboard above the chosen light position so you can check that the ceiling void above is unobstructed by joists, plumbing/heating pipework or cables, and that the joists are deep enough to accommodate the fitting. Use a semi-recessed type if they are too shallow.

2 Use compasses (or a template if one is provided with the fitting) to mark a circle on the ceiling that matches the fitting's diameter. Drill a starter hole inside the circle, and cut round the outline with a padsaw. Remove the waste.

3 Turn off the power at the consumer unit and check that the circuit is dead. Run the wiring to the new light and its switch from a four-terminal junction box connected into the main circuit as shown or on a spur from a junction box or ceiling rose (see pages 80 and 81).

4 Connect the supply cable cores to the terminal block on the light fitting. If more than one light fitting is being installed, loop the cable out of the first fitting and on to the next. If the fitting is in a poorly ventilated area or in a protective box cover the cores first with heat-resistant sleeving.

5 Screw a pad of heat-resistant material – a piece of plasterboard or a mineral-fibre ceiling tile – to the underside of the floorboard immediately above the light. In lofts, pull insulating material back from around the light and make up a box to protect it from accidental damage. Leave one side open for ventilation, and allow at least 50mm (2in) clearance all round the fitting.

6 Feed the supply cable into the ceiling void and push the fitting up into the hole in the ceiling. Most recessed lights are held in place by spring-loaded adjustable clips; make sure that these hold the fitting tightly against the ceiling surface. Then fit the decorative trim, add the lamp and restore the power.

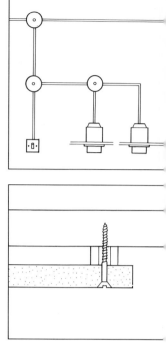

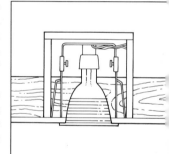

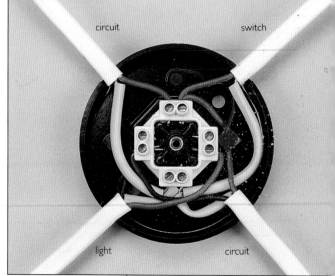

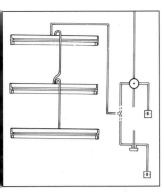

The basic construction of an illuminated ceiling is a lattice of lightweight aluminium beams and cross-bearers which are supported on edge trims fixed all round the room at the chosen ceiling level. Extra support brackets or wires may have to be used to prevent the ceiling from sagging on wide spans. The squares of the grid are then filled in with translucent panels, which must be fire-resistant in order to comply with the latest fire regulations. You can if you wish have translucent panels in just part of the ceiling and fit solid panels elsewhere – round the perimeter of the room, for example. You need a minimum of about 150mm (6in) between the old and the new ceiling surfaces to allow the light fittings to be installed.

I Turn off the power at the consumer unit and check that the circuit is dead. Disconnect the existing ceiling rose or light fitting; you will probably be able to reuse the wiring to supply the new fluorescent lights. Then paint the ceiling white to ensure the maximum reflectiveness once the suspended ceiling has been installed.

2 Install the fluorescent light fittings on the ceiling surface. The number of fittings and total tube wattage required depends on the size and type of ceiling panels chosen; the ceiling kit manufacturer will be able to advise you on this point.

3 Run the cable from one light's terminal block to another and run a cable back from the first fitting to a convenient connection point on the main lighting circuit.

4 Provide a power supply either by reconnecting the wiring at the original lighting point to a junction box and connecting in the spur cable to the new lights, or by cutting a new four-terminal box into the main lighting cable to supply the spur. See opposite. In the former case the original switch will control the new lights; in the latter, a new switch will have to be installed and wired into the new junction box. Alternatively, if it saves on cable, cut a three-terminal junction box into the circuit

FITTING AN ILLUMINATED CEILING

An illuminated ceiling is a suspended ceiling made from translucent materials and lit from above, usually by a series of fluorescent light fittings. It provides a uniform diffuse light, and is popular in kitchens and bathrooms, especially in older properties with high ceilings.

Safety Warning
The flicker from a combination of fluorescent tubes, particularly in a small room such as a bathroom, can cause epileptic fits in people prone to them. Special control gear is needed to avoid this risk, and because this is expensive, it may make more sense to light the room in another way.

THINGS YOU NEED
- Ceiling grid kit
- Fire-resistant ceiling panels
- Spirit level
- Fluorescent light fittings.
- Junction boxes
- 1mm² two-core-and-earth cable
- Green/yellow PVC earth sleeving
- Red PVC insulating tape
- Electrical tools
- General d-i-y tools

cable and run a spur to a four-terminal junction box wired for the new lights. Restore the power and test the lights.

5 Put up the ceiling grid, following the manufacturer's instruction carefully. Then drop in the panels, cut to size if necessary round the edges of the room, to complete the ceiling. The panels can be removed at any time for cleaning or for replacing fluorescent tubes or their starters when they eventually come to the end of their working lives.

Left: You can use a junction box cut into the lighting circuit cable to supply a spur for the new light fit-

FITTING EXTRA-LOW-VOLTAGE LIGHTING

Extra-low-voltage lighting, widely used in commercial premises, is now finding its way into the home too. It offers a range of exciting lighting effects and greatly reduces power consumption.

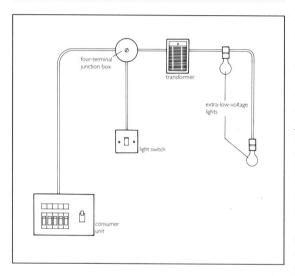

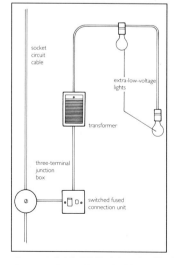

An extra-low-voltage lighting circuit can be supplied by its own circuit (top) or on a fused spur from a socket outlet circuit (bottom)

Extra-low-voltage lighting uses a transformer with an output of 12 volts to supply power to small tungsten-halogen lamps similar to those used in car headlamps. These produce up to three times as much light as a conventional lamp yet are much smaller, so the fittings containing them are also smaller and less obtrusive than mains-voltage fittings. When compared with conventional lighting, power consumption is reduced by as much as 60 per cent, and lamp life is longer too – an average of 3,000 hours, compared with about 1,000 hours for an ordinary filament lamp.

The transformer may be sited away from the fittings it supplies (typically, in a ceiling void) or may be built into the fitting if this contains two or more lamps, such as a spotlight or tracklight unit. Choose one with an output wattage to match the lamp wattage required. For example, a 50-watt transformer can supply two 20-watt lamps or one 50-watt one; a 160-watt transformer can supply up to three 50-watt lamps or eight 20-watt ones. Note that it should supply not less than 70 per cent of its stated maximum output; otherwise lamp life will be shortened. Any transformer used for extra-low-voltage lighting should comply with BS3535, and must be supplied either from a spare 5/6-amp fuseway in the consumer unit or via a spur taken off an existing circuit and protected by a 5-amp fuse. The transformer must be equipped with short-circuit protection if it is supplying more than one light fitting.

The lamps are small sealed units with multi-faceted mirrored reflectors, rather like miniature PAR spot lamps, and

give a clear white light that is also cool – useful for illuminating delicate objects such as floral displays or paintings. They should comply with BS4533. A range of different beam types is available, ranging from a thin pencil to a broader spot.

Wiring of extra-low-voltage lighting is carried out using ordinary mains-voltage cables, but larger sizes may be needed for long cable runs; the table gives details. Note that the extra-low-voltage side of the circuit between transformer and light fittings must not be earthed.

1 Choose the number and type of extra-low-voltage light fittings required to create the lighting effect you want, and select a suitable transformer to supply them.

2 Plan the positions of the fittings and the transformer to keep circuit lengths – and hence cable sizes – to a minimum. Use the table opposite to select the cable size needed to match the lamp wattage and proposed circuit length.

3 Install the light fittings at their chosen locations, ensuring that they have sufficient clear space above and around them to prevent them overheating. Then connect the live and neutral cable cores to their terminal blocks. Cut the earth core back flush with the cable sheathing; extra-low-voltage lighting circuits are not earthed. Where the cable supplies more than one light, these are wired in series with the cable running from one light to the next and terminating at the last light on the circuit.

4 Run the cable to the transformer and connect it to

the terminals. Again there is no earth connection. Ensure that the transformer is well supported if resting on a ceiling surface, and that it has adequate ventilation all round.

5 Turn off the power at the mains and check that the circuit is dead. Use 1.5mm² two-core-and-earth cable to provide the mains-voltage power supply to the transformer. Ideally it should run from a spare 5/6-amp fuseway in the consumer unit, but it can also be taken as an unfused spur from an existing light circuit, or as a fused spur from a socket circuit (see page 82). If the latter method is used, the fused connection unit should be fitted with a 5-amp fuse.

6 Provide switching for the extra-low-voltage lights by cutting in a four-terminal junction box at a convenient point on the supply cable to the

Wiring Regulations

Choose the correct cable size for the load and the length of the extra-low-voltage circuit.

50-watt transformer			
Lamp load	Cable size		
	1mm²	1.5mm²	2.5mm²
1 x 50W	4m	5.9m	9.6m
2 x 20W	5m	7.5m	12.2m
160-watt transformer			
Lamp load	Cable size		
	1.5mm²	2.5mm²	6mm²
3 x 50W	2m	3.2m	7.5m
8 x 20W	1.8m	3m	7m

transformer, and wire a switch into it as for mains-voltage wiring.

7 If the extra-low-voltage installation is replacing an existing mains-voltage ceiling rose or light fitting, remove this and reconnect its mains-voltage cables to a four-terminal junction box. The supply to the transformer takes the

place of the flex to the light, and the original switch will control the new lights.

8 Fit lamps in the light fittings, check all the connections, restore the power and test the installation.

Below: Ceiling-mounted extra-low-voltage track lighting set with integral transformer

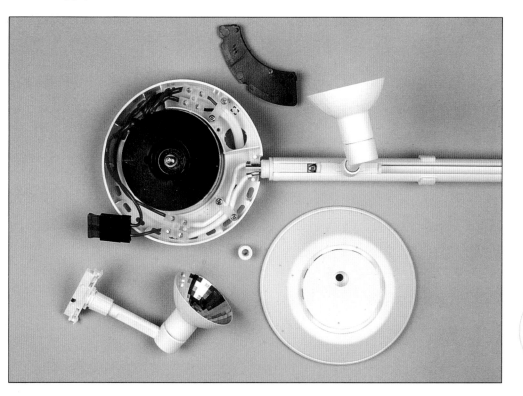

WORK ON SOCKET OUTLET CIRCUITS

As with lighting circuits, the first step to take before starting work is to establish what circuits you have and whether they are safe. If you have standard 13-amp socket outlets, find out whether they are on radial or ring circuits by checking the connections at the consumer unit – two cables run from a fuseway to a ring circuit, one to a radial. Turn each one off in turn to identify which socket outlets it supplies and make a sketch of the system. Check how much of the house each circuit serves: there are no limits to the number of outlets on the circuit, only to the floor area (see pages 42-3).

If you have old-fashioned round-pin socket outlets, the system should be assessed by a professional electrician before you do any work on it.

Much work on socket circuits will be for the sake of convenience – making minor changes to provide the number of outlets you want in the most useful locations.

Other jobs include installing a dedicated supply for permanently installed and bathroom appliances and adding RCD protection for a socket provided to supply outdoor power tools.

CONVERTING SOCKET OUTLETS

If you need more socket outlets and have mainly single outlets on your system at present, converting these to double (or even triple) outlets will give you extra capacity without the need to do any actual wiring work. However, you should not convert any old-style outlets taking round-pin plugs in this way unless the wiring has been certified safe by a professional electrician first.

Since there are no restrictions on the number of outlets a modern ring or radial circuit can supply within the specified maximum floor area, you can convert all your outlets in this way, with one exception. Any outlet which you find is wired as a spur (see pages 92-3) cannot be converted to a triple outlet, only to a double one.

Option 1:
SURFACE-MOUNTED TO SURFACE-MOUNTED

If your existing single sockets are surface-mounted and you are happy for the new outlets to be mounted in this way too, the conversion is very simple and quick to carry out. This is the most direct conversion of all the options.

1 Turn off the power at the consumer unit and check that the circuit is dead. Undo the faceplate fixing screws and ease the faceplate away from its box. Disconnect the cable cores from the terminals.

2 Unscrew the old mounting box and set it aside. Remove a knockout from the back of the new box so you can feed in the cable(s), then set your spirit level on top of the box to get it level and mark the positions of its screw holes on the wall. Drill and plug these and screw the box to the wall.

3 Reconnect the cable cores to their terminals. If any earth core is bare, cover it with a length of green/yellow earth sleeving. Then fold the cable(s) neatly back into the mounting box and attach the new faceplate.

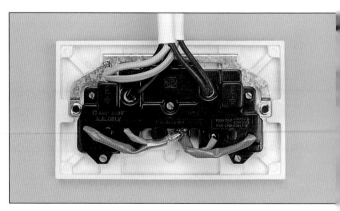

switch off first

Option 2:
FLUSH-MOUNTED TO SURFACE-MOUNTED

If the existing outlet is flush-mounted and you do not want to go to the trouble of enlarging the hole to take a larger flush box, fit a surface-mounted box over the flush one. You may not be able to do this if the existing cable tails are very short, since the new faceplate will be further away from the wall surface than the existing one. Again this is a very simple and quick conversion to carry out.

1 Turn off the power at the consumer unit and check that the circuit is dead. Unscrew the old faceplate and disconnect the cores.

2 Remove a knockout from the new box, feed in the cable(s) and attach it to the fixing lugs of the old box with the machine screws that attached the old faceplate. Alternatively, use a twin converter frame, which is fitted in the same way but has a hollow back. Reconnect the cable cores to the new faceplate as for Option 1.

re-using Single Socket Outlets

Do not throw the old single socket outlets away. You can buy special dual mounting boxes that accept two single wiring accessories side by side, and mount them in pairs to gain some additional outlets at minimal expense.

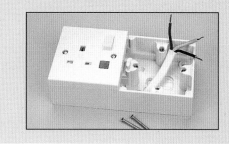

Option 3:
SURFACE-MOUNTED TO FLUSH-MOUNTED

If you want the new outlet to be flush-mounted you will have to cut a recess or hole for it in the wall without damaging the cable(s).

1 Turn off the power at the consumer unit and check that the circuit is dead. Remove the old socket and mounting box, and try to work out the direction the cable takes to the outlet position. Then hold the new flush box against the wall in a position that best avoids the cable run, and draw round its outline.

2 On solid walls, chop out the plaster and masonry to the required depth, testing the box for fit as you proceed. See page 70.

3 On plasterboard walls, check whether the old mounting box was secured to a stud. If it was, the new flush box will have to be fitted to one side of the stud or the other; you can tell which by examining how the cable entered the old box. See page 71.

4 Reconnect the cable cores to the new faceplate as in Option 1, and fit a sleeved earth link between the faceplate and box earth terminals if using a metal mounting box. Attach the faceplate to the box as before.

THINGS YOU NEED

For all options
- New double/triple socket outlet
- Green/yellow PVC earth sleeving
- General d-i-y tools

For Options 1 & 2
- New double/triple surface mounting box or twin converter frame

For Options 3 & 4
- New double/triple surface mounting box to suit wall type
- Sharp brick bolster and club hammer
- Drilling guide (optional but helpful)
- Padsaw for plasterboard stud walls
- Filler and filling knife

Option 4:
FLUSH-MOUNTED TO FLUSH-MOUNTED

If you want the new outlet to be flush-mounted like the existing one, you have to remove the old box and enlarge the hole.

1 Turn off the power at the consumer unit and check that the circuit is dead. Remove the old faceplate. In solid walls, unscrew the mounting box and prise it out. Free a hollow-wall box either by rotating the fixing lugs or by inserting a screwdriver at either side to press in the spring-loaded lugs.

2 In solid walls, decide how best to enlarge the recess; this will depend on the direction from which the cable enters. It is generally best to enlarge the recess equally at both sides. Mark the outline of the new box on the wall with a pencil, then install it as shown on page 70.

3 In hollow walls, enlarge the cut-out with your padsaw after checking nearby stud positions. Then fit the new box as shown on page 71.

4 Reconnect the new faceplate as in Option 1.

Above: Mounting boxes – choose the correct box for the option you have selected

ADDING SOCKET OUTLETS

Converting single socket outlets into double ones can give you only a limited number of new outlets, and you have no choice over their positioning. If you need additional socket outlets in positions where there are none, adding extra socket outlets to your existing socket circuits is one solution.

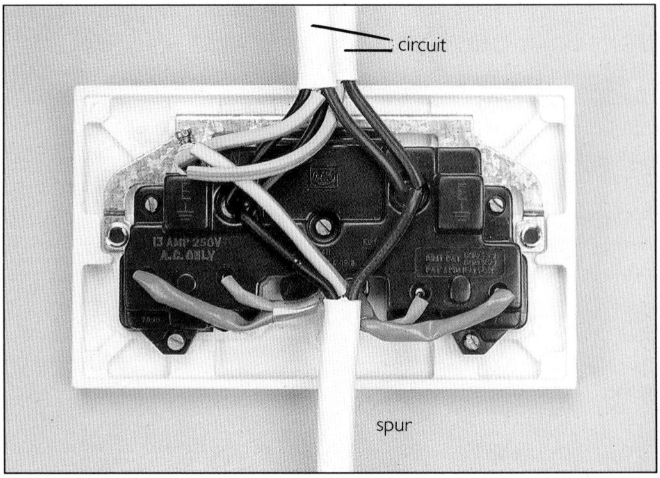

circuit

spur

THINGS YOU NEED

- New socket outlet
- New mounting box
- 2.5mm² two-core-and-earth cable
- Green/yellow PVC earth sleeving
- Electrical tools
- Continuity tester
- General d-i-y tools

Option 1:
TAKING A SPUR FROM AN EXISTING OUTLET

In most situations the simplest way of adding extra outlets is to connect a branch cable – a spur – to the terminals of an existing socket outlet, and to run the spur to the location of the new outlet. However, you cannot use any outlet for the spur connection; it must be one that is on the existing main circuit cable (see box), not one that is supplying a spur or is on a spur. Checking the status of a particular outlet involves a combination of testing and detective work.

It is sometimes possible to identify spurs by the fact that the spur cable, outlet and mounting box differ from those used elsewhere on the system. Outlets with two cables are the most difficult to identify; you may have to use a continuity tester.

Turn off the power at the consumer unit and check that the circuit is dead. Unscrew a socket that is supplied by the cable you intend to cut into. Disconnect the red and black conductors and connect a con-tinuity tester between the black cores. If the tester gives a reading the cable is part of the ring circuit; if not, it is on a spur.

Fit the mounting box at the new outlet position, and run cable from it back to the outlet you have identified as a suitable supply point. Turn off the power at the consumer unit and check that the circuit is dead. Connect the cable to the new outlet as shown, and screw it to its mounting box.

2 Double-check that the power is off, then open the 'supply' outlet by unscrewing its faceplate and easing it away from its mounting box. Feed the spur cable into the box and connect it to the terminals as shown above. Ease the three cables back into the mounting box and reattach the faceplate. Restore the power.

Wiring Regulations

The Wiring Regulations allow an unlimited number of socket outlets on each circuit; the only restriction is the floor area the circuit serves. A 30/32-amp ring circuit can serve rooms with a floor area of up to 100sq m (1075sq ft), while the limit for a 30/32-amp radial circuit is 50sq m (540sq ft) and for a 20-amp radial circuit it is 20sq m (215sq ft). If you are adding outlets as spurs and the new outlets will be in rooms separate from those served by the circuit, check that you will not exceed these limits.

*The following sockets **cannot** be used to supply a spur socket:*
- *an existing outlet with three cables, i.e. one which is already supplying a spur;*
- *an outlet with one cable on a spur connected to a ring or radial circuit;*
- *an outlet with two cables that is the intermediate socket on a two-socket spur (such spurs are not now permitted, but you may encounter them on existing installations).*

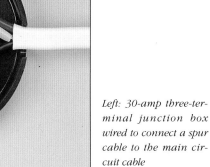

Option 2:

USING A JUNCTION BOX

The main alternative to supplying a spur from an existing outlet is to use a 30-amp three-terminal junction box to link the spur cable directly to an existing circuit cable. Use a round box if making the connection in an underfloor void, or a white rectangular box if you have surface-mounted wiring.

1 You must first check that the cable you plan to use is not supplying a spur. Open up a socket supplied by the cable and check as in Option 1.

Left: 30-amp three-terminal junction box wired to connect a spur cable to the main circuit cable

2 Fit the new mounting box and outlet as described in Option 1, and run the spur cable back to your chosen connection point.

3 With the power off, screw the base of the junction box into place at the connection point, and connect the cores of the cut circuit cable and the spur cable to it as shown. Replace the box cover and restore the power.

Left: Rectanglar surface-mounted junction box wired to supply a spur cable where the wiring is surface-mounted

Option 3:

WIRING DIRECTLY INTO A CIRCUIT

1 If the existing circuit cable runs beneath floorboards close to where you want an additional outlet, is not clipped to the joists and has enough slack, you may be able to draw a loop of cable up behind the skirting board.

2 Fit the new mounting box, and feed in the cable loop. With the power off, cut it and connect the cores to the new outlet faceplate as shown. Ease the cables back into the box, replace the faceplate and restore the power.

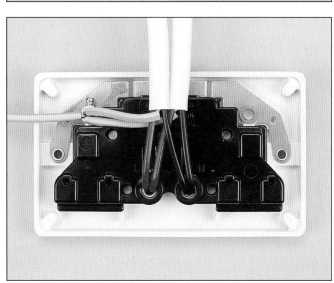

Left: Socket outlet added into the circuit. If using a metal mounting box add an earth link as shown to the box earth terminal

REORGANISING CIRCUITS

Another way to provide additional socket outlets is to reorganise your existing circuits. You may also have to do this if, for example, you extend your home and you find that the new floor areas being served will exceed the limits set by the Wiring Regulations for ring and radial circuits.

There are three ways to reorganise circuits serving socket outlets to gain extra capacity without the need for full-scale rewiring. The first is to extend a ring circuit to serve a larger area, by breaking into the ring and wiring in a further 'loop' of cable between the cut ends of the original circuit.

The second is to split an existing ring circuit into two halves, effectively creating two smaller ring circuits which can themselves then be extended to serve larger floor areas; however, you need a spare 30/32-amp fuseway at the consumer unit in order to be able to do this.

The third involves converting an existing radial circuit into a ring circuit by running cable from the last socket on the circuit back to the consumer unit via a number of additional socket outlets.

THINGS YOU NEED

- New socket outlets and mounting boxes
- 2.5mm² two-core-and-earth cable
- Green/yellow PVC earth sleeving
- Electrical tools
- Continuity tester
- General d-i-y tools

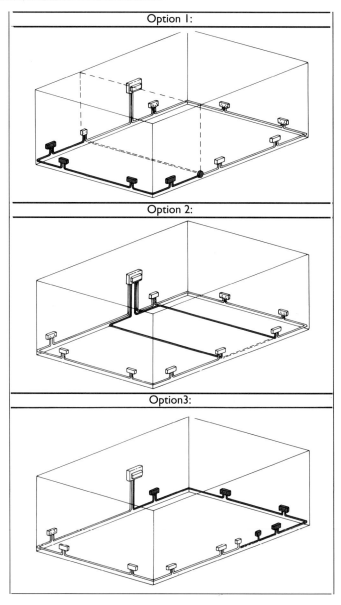

Option 1:

Option 2:

Option3:

Option 1:
EXTENDING A RING CIRCUIT

1 Start by planning the positions of the new socket outlets, remembering that some of them can be supplied as spurs if this makes for more economical use of cable. This will enable you to plan the optimum cable route for the new 'loop', and will also dictate where best to connect it into the existing circuit. Check that the floor area served by the extended ring will not exceed 100sq m (1075sq ft).

2 Install the new socket outlets and run cable from them back to the chosen connection points – either at existing socket outlets or via new 30-amp junction boxes. Turn the power off at the consumer unit and check that the circuit is dead. Breach the ring by disconnecting the cable linking the two connection points, and connect in the ends of the new loop.

Option 2:

CREATING TWO RING CIRCUITS

1 As with Option 1, begin by planning where you want additional socket outlets so you can decide how to run the new cables and where best to sever the ring.

2 Turn the power off at the consumer unit and check that the circuit is dead. Disconnect one section of cable and run new cables back from the disconnection points to the consumer unit via the new socket outlets.

3 Disconnect one live core from the fuse or MCB serving the original circuit, and connect it to a spare/new fuse/MCB. Then use a continuity tester to identify which new live cable core completes which circuit, and connect the new live core to the appropriate fuse or MCB. Take the new neutral and earth cores to their respective terminal blocks.

Option 3:

'RINGING' A RADIAL CIRCUIT

1 Turn the power off at the consumer unit and check that the circuit is dead. Locate the last socket outlet on the circuit and run new cable on from it to the new outlet positions and back to the consumer unit.

2 Connect the live core of the new cable to the existing circuit fuse or MCB, and the neutral and earth to their respective terminal blocks. If the existing circuit was protected by a 20-amp fuse or MCB, replace it with one rated at 30 or 32 amps.

If an existing socket will be blocked off by furniture, it can be used as a junction box to supply a spur to a new socket fitted in a more convenient position nearby. It can easily be reinstated as a socket outlet in future, simply by reversing the procedure outlined here.

1 Turn off the power at the mains and check that the circuit is dead. Unscrew the socket outlet faceplate and disconnect the cable cores.

2 Decide on the position of the new outlet, and work out the best way of running the spur cable to it from the original one. The least disruptive solution is to surface-mount it along the top edge of the skirting board for now, and to conceal it when you next redecorate the room. If you do not need the outlet, but could use an extra one in the adjacent room, consider running the spur cable directly through the wall behind the outlet and fitting another back-to-back with it.

3 Install the new mounting box and run the cable to it from the original outlet posi-

tion. Connect the cable to the old outlet faceplate (or a new one if it was a single and you could use a double one at the new location) and fit this to the new box.

4 At the original outlet position, connect the existing cable cores to the spur cable using strip connectors. Link like cores to like cores, and if the box is metal, add an earth link from the strip connector earth terminal to the terminal in the box.

Fold the cables and the strip connector neatly back into the box and fit a blanking-off plate to cover it.

RELOCATING SOCKET OUTLETS

It is commonplace to find that other people's idea of the perfect location for a socket outlet does not coincide with yours. For example, you may want to stand a large piece of furniture in front of it, or to install built-in furniture at that point, so preventing access to the socket. The solution is to relocate it.

THINGS YOU NEED

- Screwdriver
- Terminal screwdriver
- Blanking-off plate for original socket
- Strip connectors
- 2.5mm² two-core-and-earth cable
- Green/yellow PVC earth sleeving
- New socket outlet
- Mounting box
- Electrical tools
- General d-i-y tools

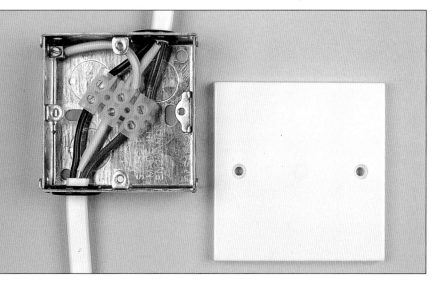

USING FUSED CONNECTION UNITS (FCUs)

As the name implies, a fused connection unit (FCU) allows individual appliances rated at up to 3kW to be permanently connected to a ring or radial circuit, instead of being plugged into a socket outlet. There are several types, each with its own specific uses.

THINGS YOU NEED

- FCU and mounting box
- Cable rated to match installation
- Green/yellow PVC earth sleeving
- Fuse to match appliance or sub-circuit
- Electrical tools
- General d-i-y tools

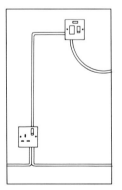

Above: Fused connection unit supplied as a spur from a ring or radial circuit

The advantage of using an FCU instead of a socket outlet for certain appliances is that it provides a dedicated connection point for that appliance. It also looks neater than a plug and socket outlet, yet can still provide localised switch control and offers the same fuse protection as a plug.

An FCU can be wired in on the main circuit, just like a socket outlet, or can be supplied as a spur. Unswitched FCUs are also used in bathrooms, where socket outlets are not allowed, to power things like extractor fans and wall heaters with their own on-off cord-pull switches. They can also supply sub-circuits to wall lights too as an alternative to wiring these into a lighting circuit.

FCUs are the same size as a single socket outlet and are available either switched or unswitched, with or without neon on-off indicators and with flex entry through the front or the edge of the faceplate.

There are also versions without the flex entry, used for wiring fused spurs where the load is fed by cable rather than flex. Switched FCUs should have a double-pole switch; check when you buy.

1 Decide where the FCU is to be installed, unless it is replacing an existing single socket outlet. For floor-standing or fixed appliances, site it so that the switch is accessible yet the flex is short and out of harm's way.

2 Turn off the power at the consumer unit and check that the circuit is dead. If you are replacing an existing socket outlet, unscrew and disconnect the socket outlet faceplate. Reconnect the live and neutral cable cores to the FCU faceplate, using the terminals marked FEED. Connect the earth core(s) – plus the earth link if the mounting box is metal – to the FCU's earth terminal.

Below: Two FCUs with flex running on to the appliance

3 If you are installing the FCU as a spur, fit its mounting box at the chosen point and run 2.5mm² two-core-and-earth cable back from it to the point where it will be connected into the main circuit – at an existing socket outlet or via a 30-amp junction box. Connect the cable to the FCU as in Step 2.

4 If the appliance flex is being connected directly into the FCU, feed the flex through the hole in the faceplate and attach its live and neutral cores to the terminals marked LOAD, and its earth core to the earth terminal. Secure its sheath in the cord grip, fold the cables neatly back into the box and screw the faceplate to the box. Check that the FCU fuse matches the appliance's wattage.

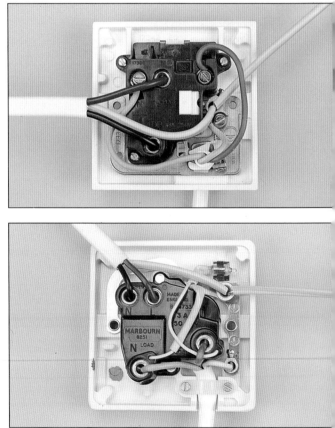

5 If the FCU is supplying a sub-circuit – to a wall light, for example, or to a heated towel rail or an extractor fan in a bathroom – cable is used for the next part of the circuit, and its rating is chosen to match the circuit requirements. For example, 1mm² cable is used to supply a wall light or fan, and 2.5mm² for a wall heater or an unswitched towel rail (the latter via a ceiling switch and a wiring accessory called a flex outlet plate, which connects cable to flex; see page 100 for more details). Run the new sub-circuit cable between the FCU and the light, appliance or flex outlet plate as appropriate. Connect the sub-circuit cable to the FCU terminals marked LOAD, and fit a fuse of the appropriate rating for the sub-circuit or appliance being served.

6 Make the final connection to the mains supply at an appropriate socket outlet or junction box.

CLOCK CONNECTORS

A variation on the standard FCU is a wiring accessory called a clock connector, which as its name implies was originally designed to supply mains-powered clocks. The flex is connected to a special square plug containing a 2-amp fuse; this fits flush within the faceplate and is retained by a small captive screw. The advantage of a clock connector over an FCU is that the appliance can be disconnected from the mains if necessary without removing the faceplate, as is the case with ordinary FCUs. It can be used to supply any appliance rated at up to 480 watts.

Below left: FCU with cable supplying the appliance (top) and flex outlet plate (bottom)

Below: Clock connector

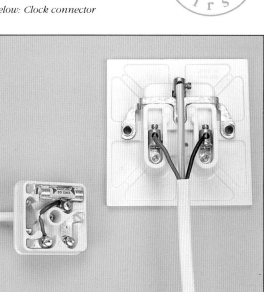

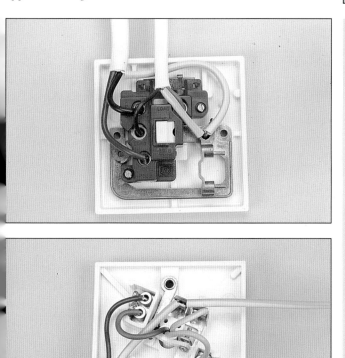

Appliance Switching

Where freestanding but enclosed appliances such as fridges, washing machines and dishwashers are supplied via a socket outlet or FCU at the back of their recesses, you have to pull the appliance right out to switch it off for repairs or servicing. To provide more convenient switching, wire the socket outlet/FCU as a spur run via a double-pole isolating switch situated above worktop level. These switches are available with a range of different appliance names stamped on the faceplate to identify what each one controls.

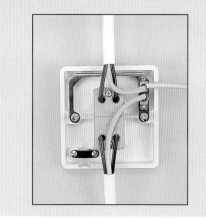

WIRING FANS AND COOKER HOODS

Extractor fans are an excellent way of controlling condensation, especially in kitchens and bathrooms (where they are now mandatory in all new homes). Cooker hoods that extract air rather than just filtering and recirculating it do the same job. Both should be permanently connected to the mains supply, usually via an FCU.

THINGS YOU NEED

- Extractor fan or cooker hood
- FCU and mounting box
- 1mm² two-core-and-earth cable
- Green/yellow PVC earth sleeving
- 3-terminal junction box for wiring via a lighting circuit
- Electrical tools
- General d-i-y tools

How you wire up an extractor fan or cooker hood depends chiefly on how conveniently you can obtain a power supply from an existing circuit, and on whether the appliance has its own on-off switch. There are several options.

Below: The spur cable from the socket outlet or junction box is connected to the FEED terminals of the FCU, the supply cable to the appliance to the LOAD terminals

Option 1:
FUSED SPUR FROM A POWER CIRCUIT

Supply the fan or hood via a fused spur connected to the circuit at a socket outlet or 30-amp junction box. Use 2.5mm² cable between the circuit and the FCU, and 1mm² cable for the supply to the fan. Fit a 3-amp fuse in the FCU. If the fan/hood does not have its own switch, site the FCU where its switch can be conveniently reached. Note that in bathrooms you must either site the switched FCU outside the bathroom or else use an unswitched FCU and run the cable to the fan via a ceiling-operated switch within the room.

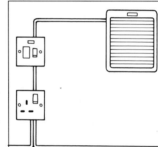

Option 2:
SPUR FROM A CEILING ROSE

Connect in a 1mm² spur cable for the fan at an existing loop-in ceiling rose if there is room for an additional cable. Use the loop-in, neutral and earth terminals. Run it to the FCU (which here acts as the fan's on-off switch) and on to the fan. See Option 1 for restrictions in bathrooms.

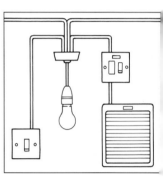

Option 3:
SPUR FROM A LIGHTING CIRCUIT JUNCTION BOX

Connect in a 1mm² spur cable for the fan at an existing four-terminal junction box if there is room for the connections. Wire the spur as in Option 2, making sure you connect it to the circuit cables, not the switch cable. Again see Option 1 for wiring restrictions in bathrooms.

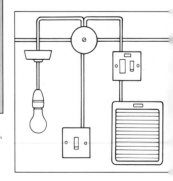

Option 4:

SPUR FROM A NEW JUNCTION BOX

If there is no room to connect in a spur cable at an existing loop-in rose or junction box, cut into the lighting circuit at a convenient point and fit a three-terminal junction box. Make sure it is a circuit cable and not a switch cable. Run the 1mm² spur cable from this to the FCU and fan as before. Again see Option 1 for wiring restrictions in bathrooms.

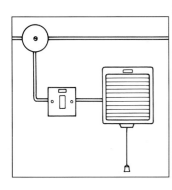

Simultaneous Operation

If you have an enclosed windowless bathroom or WC that depends on the fan for ventilation, it is best to loop in the power supply to the fan at the light so that the fan comes on as soon as the light switch is operated. Choose a fan with a built-in over-run timer; this allows the fan to continue running for a pre-set period after the light has been switched off, so guaranteeing thorough ventilation of the room.

How you wire up a WDU depends on whether the unit is supplied with its own switch or not, and on whether you are prepared to reach into the cupboard beneath the sink or would prefer a wall-mounted switch. Some more sophisticated units do not require a switch; they are activated automatically when they detect a load and the presence of running water.

1 The simplest way of wiring up a WDU is to run a spur cable to a switched FCU surface-mounted on the side wall of the cupboard housing the WDU, and then to wire the flex directly into the FCU.

2 If you want switch control above worktop level, run the spur to an FCU sited conveniently near the sink, but not so close that it will be splashed, and just above worktop level. Then run 1mm² cable down the wall into the cupboard beneath and wire it into a flex outlet plate, to which the WDU flex is then connected.

3 If the unit is provided with a wall-mounted on-off switch, check whether it is a double-pole switch. If it is, run the spur to an unswitched FCU above worktop level, then run cable on to the switch, which can be mounted next to the FCU. From there, run cable to a flex outlet plate as in Step 2. If the switch is not a double-pole type, use the arrangement described above but substitute a switched FCU for the unswitched one. (This can be used for isolating the appliance, though the manufacturer's switch should be used for day-to-day control of the unit.) Fit a 3-amp fuse in the FCU, whichever wiring arrangement is selected.

WIRING WASTE DISPOSAL UNITS

Waste disposal units (WDUs) have become increasingly popular as a quick and hygienic kitchen appliance. A WDU is basically a macerator. The motor drives grinding blades which reduce waste matter to a slurry; this is then washed away down the drain by running water through the unit. The motor is protected by an automatic cut-out to prevent overloading or to shut it off if the blades become jammed.

4 Turn the power off at the consumer unit and check that the circuit is dead. Connect the spur cable to a socket outlet that is not already supplying a spur (see pages 92-3).

(see pages 92-3)

THINGS YOU NEED

- Waste disposal unit
- Switched FCU or
- Unswitched FCU and double-pole switch
- Flex outlet plate
- Mounting boxes
- 2.5mm² two-core-and-earth cable
- 1mm² two-core-and-earth cable
- Green/yellow PVC earth sleeving
- Electrical tools
- General d-i-y tools

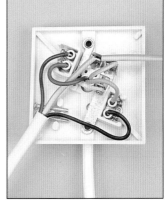

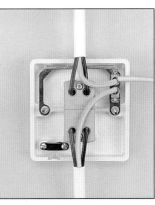

Above left: Flex outlet plate

Left: Double-pole isolating switch

WIRING WALL HEATERS AND TOWEL RAILS

Most portable electric heaters are designed to be plugged into a socket outlet, but for a wall-mounted heater a permanent connection to the mains supply is more appropriate. Heaters in bathrooms, including electric towel rails, must be permanently connected to comply with the Wiring Regulations.

THINGS YOU NEED

- Heater/towel rail
- 2.5mm² two-core-and-earth cable
- Green/yellow PVC earth sleeving
- Switched FCU with neon indicator and flex entry for wall-mounted heater
- Unswitched FCU, 15/16-amp double-pole ceiling switch with neon indicator and flex outlet plate or
- Switched FCU and flex outlet plate for towel rail
- Mounting boxes for above
- Electrical tools
- General d-i-y tools

Wall-mounted heaters cannot conveniently be plugged into a socket outlet, unless this is above the conventional skirting-board level. Towel rails in bathrooms cannot be plugged into a socket outlet at all, for the obvious reason that socket outlets are not allowed in bathrooms.

Below: Flex connections to a towel rail. Bottom: 15/16-amp double-pole ceiling switch

WIRING FOR OCCASIONAL USE

This wiring arrangement is not suitable for bathrooms or for heaters in regular use.

1 Mount a switched FCU fitted with a 13-amp fuse close to the heater position, and link the two with the heater flex. Wire the live and neutral flex cores into its LOAD terminals, and the earth core to the neutral core.

2 Connect 2.5mm² cable to the FEED and earth terminals of the FCU, and run it back to a convenient connection point on a nearby power circuit. Turn off the power at the consumer unit and check that the circuit is dead. Make the connection.

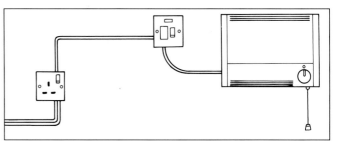

WIRING HEATERS IN BATHROOMS

1 In bathrooms, heaters such as towel rails must either be switched from outside the room or have a cord-operated ceiling-mounted isolating switch. Mount a flex outlet plate next to the heater and connect its flex to it. Then run 2.5mm² cable back from the flex outlet plate to either a switched FCU outside the bathroom, or to a 15/16-amp ceiling switch within the room and then to an unswitched FCU.

2 Connect 2.5mm² cable to the FEED and earth terminals of the FCU, and run the cable back to a convenient connection point on a nearby power circuit. Turn off the power at the consumer unit and check that the circuit is dead. Make the connection.

3 The metallic parts of a heated towel rail must be included in the equipotential bonding in the bathroom (see pages 66-7). Run a 4mm² earth cable covered in green/yellow sleeving from the earth terminal in the flex outlet plate supplying the towel rail to the nearest earth bonding connection or attach it to the cold water pipe with an earth clamp. Make sure that all the metallic parts in the bathroom are properly cross bonded to earth.

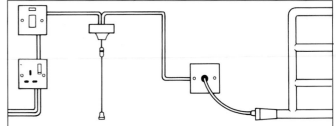

The two most convenient sites for shaver supply points are in the bathroom and the bedroom. Two types of equipment are available: the shaver supply unit and the shaver socket outlet.

Option 1:
SUPPLY FROM A LIGHTING CIRCUIT

Turn off the power at the consumer unit and check that the circuit is dead. Fit the mounting box and run 1mm² cable from it to the chosen connection point on the circuit. This can be at a loop-in rose or a four-terminal junction box, so long as there is room at the accessory for an extra cable to be connected. Otherwise cut the circuit cable (make sure it is not a switch cable) at a con-

venient point and wire the cable as a spur using a three-terminal junction box.See Options 2–4 on pages 98–9.

Make certain that the shaver supply unit is earthed to the ceiling rose or junction box.

Option 2:
SUPPLY FROM A SOCKET OUTLET CIRCUIT

Turn off the power at the consumer unit and check that the circuit is dead. Fit the mounting box and run 1mm² cable back towards the chosen connection point as in Option 1. Wire it into a fused connection unit (FCU) fitted with a 3-amp fuse, and then link the FCU to the main circuit using 2.5mm² cable, either at an eligible socket outlet or using a 30-

WIRING SHAVER SUPPLY POINTS

The widespread popularity of electric razors means that a dedicated shaver supply point is now an essential part of every new home's wiring system. Adding one to an existing system is straightforward, but be sure to choose the correct type of supply equipment for the location.

amp three-terminal junction box cut into the circuit wiring otherwise. See Option 1 on page 98.

Make sure that the shaver supply unit/socket outlet is earthed back to the circuit from which the supply is taken.

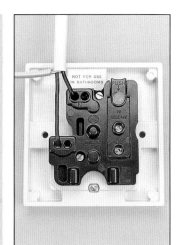

THINGS YOU NEED

- Shaver supply unit or shaver socket outlet
- Mounting box
- Two-core-and-earth cable to suit sub-circuit supplying shaver point
- Green/yellow PVC earth sleeving
- Fused connection unit
- Junction box
- Electrical tools
- General d-i-y tools

Wiring Regulations

*The **shaver supply unit** contains an isolating transformer that provides an earth-free supply to the razor. This means that the user is electrically isolated from the mains supply, and is why a shaver supply unit is the only type of socket outlet allowed in a bathroom or cloakroom. The unit contains a cut-out to prevent any other appliance being supplied from it, and it may be combined with a striplight. All such units must comply with British Standard BS3535 (and with BS4533 if combined with a striplight).*

*The smaller (and cheaper) **shaver socket outlet** can be used in bedrooms and other places safely away from a water supply. This does not contain a transformer, but is fitted with a self-resetting overload device protected by a 1-amp fuse. It should conform to BS4573.*

Cross-bonding Striplights

If installed in a room containing a fixed bath or shower, combined shaver supply units and striplights must be connected to the equipotential bonding in the bathroom with 4mm² cable. See Wiring heaters in bathrooms opposite and pages 66-7.

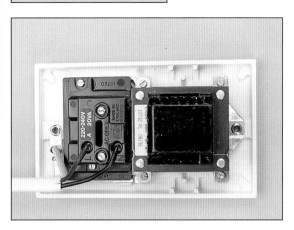

Top: *Shaver socket outlet*

Bottom: *Shaver supply unit*

WIRING HEATING CONTROLS

Modern central heating systems are fitted with a range of controls which ultimately take their power supply from the mains. Although the wiring details of individual controls will vary from make to make, the basic connections are broadly similar.

THINGS YOU NEED

- Controls to suit system
- Fused connection unit
- Mounting box
- Multi-terminal junction box (if needed)
- 1mm² two-core-and-earth cable
- Green/yellow PVC earth sleeving
- Multi-core flex to interconnect controls
- Flex outlet plate and mounting box for connections at motorised valves
- Electrical tools
- General d-i-y tools

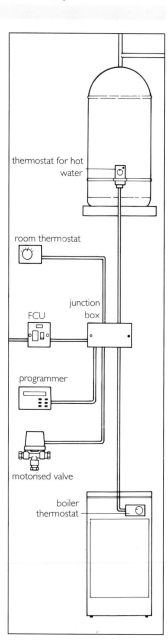

thermostat for hot water

room thermostat

FCU

junction box

programmer

motorised valve

boiler thermostat

The heating system's main control is the programmer, which governs when space or water heating (or both) are supplied. The room thermostat detects air temperature changes and switches the space heating on or off accordingly. The cylinder stat monitors the temperature of stored water in the hot cylinder, and activates a motorised valve when necessary to divert water to the cylinder's heat exchanger.

All these components are usually wired either into the programmer or into a multi-terminal junction box; whichever is used it takes its power supply from a fused connection unit on the socket outlet circuit or on a spur taken off it.

On sophisticated systems there may be additional motorised valves used to divide the heating into zones, and extra controls such as frost stats designed to override other controls and switch the heating on when the external temperature drops.

Left: The various controls are usually wired to a multi-terminal junction box, which is connected to a power circuit via a fused connection unit (top of page)

1 Position the programmer in the most convenient place; next to the boiler if this is indoors, or in the lounge, kitchen or hall.

2 Position the room thermostat out of draughts and away from heat sources. If placed in the main living room, it will keep that room at the pre-set temperature.

3 Mount the cylinder stat on the hot cylinder, after removing or cutting back any insulation to allow contact with its bare metal wall. Connect in motorised valves as required by the system design.

4 Link the controls to the programmer or junction box as dictated by their wiring diagrams.

5 Turn off the power at the consumer unit and check that the circuit is dead. Run the power supply for the controls in 1mm² cable, taken from a fused connection unit fitted with a 3-amp fuse.

You may decide you would prefer to have RCD protection for all your socket outlets (see page 118 for more details). However, you can provide an equivalent level of protection at one designated socket outlet position by fitting an RCD socket outlet. You can either replace an existing double outlet with the RCD outlet, or wire up the new outlet as a spur off an existing circuit. Better still, you can install a special weatherproof socket outlet on the outside wall of the house and avoid having extension flexes trailing through windows and doorways.

RCD socket outlets designed for indoor installation fit a double mounting box 25mm or 35mm deep depending on the design, and provide just one outlet. If you are fitting one to an existing double box, check the available depth before buying the RCD outlet. Those for outdoor use come complete with a weatherproof enclosure, and are available in one-gang and two-gang versions. With outdoor versions you usually need to use a special weatherproof plug.

FITTING AN RCD SOCKET OUTLET

On new wiring installations, any socket outlet supplying an electrical appliance, d-i-y or garden power tool that is being used out of doors must be protected by a high-sensitivity residual current device (RCD) to comply with the current Wiring Regulations. On existing installations without this level of protection, it is an excellent safeguard to fit a socket outlet which contains an RCD for powering electrical equipment when it is used out of doors.

Option 1:

REPLACING AN EXISTING SOCKET OUTLET

1 Turn off the power at the consumer unit and check that the circuit is dead. Unscrew the faceplate of the double socket outlet and disconnect the cable cores. Check that the box is deep enough to accept the new RCD socket outlet.

2 Connect the cable cores to the RCD outlet faceplate, adding an earth link between its earth terminal and the one in the mounting box if this is metal.

3 Fold the cables back into the box and screw on the faceplate. Restore the power, then press the test button to check the operation of the RCD, and reset it ready for use.

Option 2:

RUNNING A SPUR TO THE NEW OUTLET

1 Install the mounting box or enclosure and connect cable to the RCD outlet as in Option 1.

2 Turn off the power at the consumer unit and check that the circuit is dead. Run the cable from the new outlet position to a suitable connection point on a nearby socket outlet circuit (see pages 92-3 for more details).

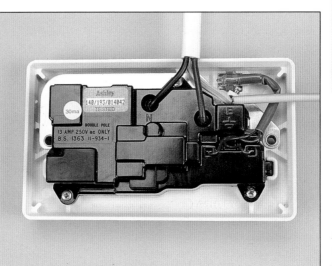

THINGS YOU NEED

- RCD socket outlet
- Mounting box (if required)
- 2.5mm² two-core-and-earth cable plus earth sleeving for new installations
- Electrical tools
- General d-i-y tools

Left: RCD socket outlet wired on a spur from a socket outlet circuit

WORK ON OTHER CIRCUITS

A modern house wiring system is likely to contain several other circuits apart from those supplying lights and socket outlets. These may include separate circuits to an immersion heater, to a cooker or separate oven and hob, to an instantaneous shower, to an outbuilding, to electric storage heaters and perhaps also to mains-operated doorbells, a burglar alarm system and even a series of mains-powered smoke detectors.

WIRING IMMERSION HEATERS

Even if the household's water is heated by a central heating system, it is usual for an electric immersion heater to be installed in the hot water cylinder. This can be used during the summer when it is uneconomical to run the boiler simply to provide hot water, and is also useful for boosting the water temperature during periods of high demand for hot water. In homes without a boiler or a central heating system, it may be the only source of hot water unless a multi-point water heater is installed instead.

The commonest type of immersion heater has a single element which is fitted into a special boss in the dome of the hot cylinder. Dual-element types have one long and one short element, controlled by separate switches or by a combined 'dual' switch; this allows either just the top of the tank or its entire contents to be heated up, depending on the likely water demand. A similar facility can be provided by fitting two separate short elements into the side of the cylinder, one near the top and the other near the base.

The heater must have its own circuit, wired from a 15/16- or 20-amp fuse or MCB using 2.5mm² cable. This terminates close to the heater position at a 20-amp double-pole switch and flex outlet (these are available marked 'WATER HEATER' on the faceplate for use when the switch is remote from the heater). The heater is wired to the switch with 1.5mm² three-core heat-resisting flex. The flex can be looped in and out of a timer to give automatic water heating at pre-set times of the day.

If the house has a separate off-peak electricity supply such as Economy 7 the heater can be wired from that consumer unit so water is heated with cheaper electricity, but obviously with this arrangement the temperature cannot be topped up during the day. A better arrangement is to have dual or twin elements, with night-rate electricity supplying the longer/lower element and day-rate power supplying the shorter/upper one. Contact your local electricity supplier for details of their off-peak tariffs before installing any such circuits.

Alternative Switching Arrangements

If the heater is situated within a bathroom or shower room and it or its switch are within reach (2.5m; 8ft 3in) of anyone using the bath or shower, you must either fit the switch outside the bathroom or use a ceiling-mounted cord-operated switch instead. In either case, the cable then runs on to a 20-amp flex outlet plate close to the heater, from where the flex runs to the element. When buying components for this arrangement, check that the flex outlet plate can cope with the thick sheath of the heat-resisting flex; not all makes can.

You can provide two-way switching for an immersion heater if you wish, with one switch by the heater and the other in the kitchen, for example. To do this you need a selection of special modular wiring accessories that can be fitted to a mounting grid. For the switch next to the heater you need a four-gang grid containing a 20-amp double-pole switch, a 20-amp two-way single-pole switch, a flex outlet and a neon indicator; for the remote switch you need another two-way single-pole switch and indicator, fitted to a two-gang grid. Two-core-and-earth cable links the indicators and 2.5mm² three-core-and-earth cable links the two-way switches.

Right: 20-amp double-pole switch

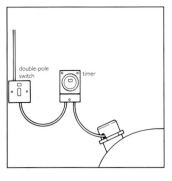

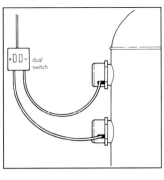

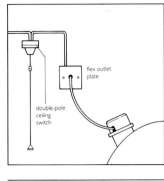

THINGS YOU NEED

- 20-amp double-pole switch and mounting box *or*
- 15/16-amp double-pole ceiling switch and surface or conduit box *or*
- Dual switch and mounting box (for dual and twin elements)
- Flex outlet plate and mounting box (for remote switching arrangements only)
- 2.5mm² two-core-and-earth cable
- 1.5mm² three-core heat-resisting flex
- Timer if required
- Electrical tools
- General d-i-y tools

Option 1:
STANDARD SWITCH INSTALLATION

1 Mount the 20-amp switch within the airing cupboard, as close to the heater boss as is practicable. Connect the cores of the flex to its LOAD and earth terminals, and secure the flex in the cord grip.

2 Connect the cores at the other end of the flex to the heater terminals (via a timer if one is being used). The live core goes to the empty terminal on the heater thermostat, while the neutral and earth cores go to their respective terminals. Check that a live link is present, connecting the other thermostat terminal to the heater's live terminal.

3 Run 2.5mm² cable from the FEED and earth terminals of the double-pole switch back to the consumer unit, turn the power off at the consumer unit and connect the live core to a 15/16 or 20-amp fuseway or MCB. Take the neutral and earth cores to their respective terminal blocks.

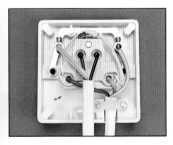

Option 2:
USING A DUAL SWITCH

1 If you are using a dual switch to control dual or twin elements, connect the live flex cores to the adjacent strapping terminals of the two-way switch, wire both the neutral cores into the neutral LOAD terminal of the double-pole switch, and take the earth cores to the earth terminal.

2 Connect one live flex core to each thermostat terminal, and the neutral and earth cores to their own terminals.

3 Connect the cable cores to the FEED and earth terminals of the double-pole switch. Fit a short live link between the vacant live FEED terminal of the double-pole switch and the common terminal of the two-way switch. Run the circuit cable back from the FEED terminals to the consumer unit as in Option 1 Step 3. Check that you have wired the elements correctly by turning the main switch on and setting the second switch to SINK. If water at the bottom of the cylinder gets hot, turn the power off at the consumer unit and reverse the live flex core positions.

Right, clockwise from top: 20-amp double-pole ceiling switch, flex outlet plate, dual switch

---

Option 3:
REMOTE SWITCHING

1 Mount the flex outlet plate close to the heater position, wire in the flex and secure its sheath in the cord grip. Connect its other end to the heater as in Option 1 Step 2.

2 Run 2.5mm² cable from the flex outlet plate to the switch – either the same type as was used in Option 1 Step 1 but mounted outside the bathroom, or a ceiling-mounted cord-operated switch within the room. Connect it to the LOAD and earth terminals. Then run the rest of the circuit as in Option 1 Step 3.

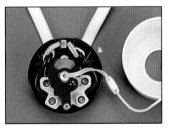

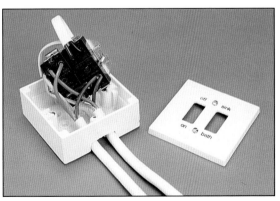

WIRING ELECTRIC COOKERS

Cookers, both free-standing types and built-in ovens and hobs, use a lot of electricity and so always have their own separate circuit wired from a 30- or 45-amp fuseway or MCB in the consumer unit.

It has over the years been common practice to supply cookers via a wiring accessory called a cooker control unit, which combines the DP isolating switch with a single 13-amp socket outlet. If the unit is positioned above the cooker or hob position, rather than to one side of it as it should be, there is a risk that any flex plugged into the socket outlet could accidentally trail across a hot hotplate, with highly dangerous consequences.

For this reason it is better to ensure that the worktop area is adequately supplied with power by ordinary socket outlets and to use just a DP isolating switch of the appropriate current rating for the cooker. If you do want to use a cooker control unit install it at least 600mm (2ft) to the side of the cooker.

THINGS YOU NEED

Switch ratings and cable sizes depend on the wattage of the cooker – see box

- Double-pole switch(es)
- Surface or flush mounting box(es)
- Cooker connection unit and flush mounting box (free-standing cookers)
- Cable
- Green/yellow PVC earth sleeving
- Electrical tools
- General d-i-y tools

Top: wiring for a free-standing cooker. Below: options for wiring separate built-in oven and hob using one or two switches depending on the position of the components

The circuit cable runs first to a suitably rated double-pole (DP) isolating switch close to the cooker position. If a free-standing cooker is fitted, more cable then runs on to a point low down at the back of the cooker recess, where it is wired into a special cable outlet known as a cooker connection unit. The cooker is connected into this via a length of the same cable as is used for the rest of the circuit – the only occasion where cable is used rather than flex to link a free-standing appliance to the mains supply. The cable should be about 2m (6ft 6in) long to allow the cooker to be pulled right out of its recess for cleaning or maintenance.

If the oven and hob are separate built-in components, cable from the switch is connected directly to the terminals of each component. It either runs first to one and then on to the other, or separate cables are run to each one from the switch. Both components must be within 2m of the switch; if this requirement cannot be met due to the layout of the kitchen, two 30/32-amp switches must be used, each within 2m of the component it controls. The power supply to the second switch and component is wired from the first switch using the same size of cable.

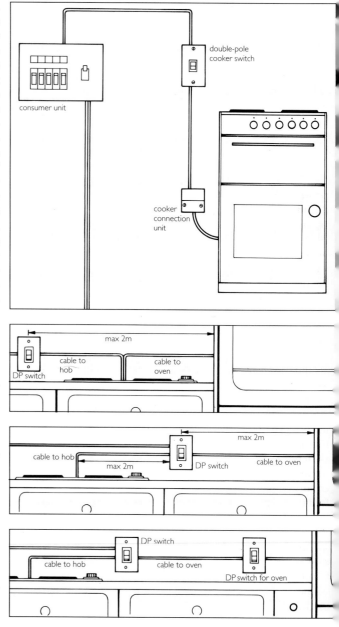

Option 1:
FREE-STANDING COOKERS

1 On the rear wall of the cooker recess, install the cooker connection unit on a flush mounting box. Run cable from its terminals back to the position of the DP switch, which must be within 2m (6ft 6in) of the cooker, and connect the cores to the LOAD terminals.

2 Connect the circuit cable cores to the switch FEED terminals, and run it back from the switch to the consumer unit (or to a new unit if no spare fuseway or MCB exists; see page 119 for more details). Turn off the power at the consumer unit and connect the live core to the fuseway or

MCB; take the neutral and earth cores to their respective terminal blocks.

3 With the power still off, fit a 2m length of cable between the cooker connection unit and the cooker terminals. Slide the cooker into its recess and restore the power.

Left: Wiring at the double-pole cooker switch for a free-standing cooker or for built in components wired one after the other (top two diagrams opposite)

Below left: Cooker connection unit

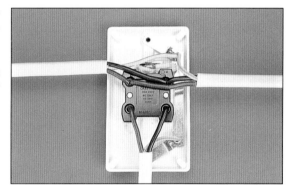

Below: Wiring at the cooker switch for separate built-in oven and hob. The cables either run on to the individual components directly, or supply one of them via a second switch wired as above (bottom two diagrams opposite)

Option 2:
BUILT-IN COMPONENTS

1 Decide whether the components can be controlled by a single switch, or whether each needs its own switch. Then plan the optimum cable arrangement, wiring the switch(es) and components in series or in parallel as appropriate.

2 Fit the switch mounting box(es) and run cable to each component. Make the connections to the components' terminal blocks and to the LOAD terminals of the switch(es). Slide the components into their cabinets, ensuring that the cable can fold up neatly as you do so.

3 If only one switch is being used, connect the supply side of the circuit to its FEED terminals. If two are needed, connect the supply cable cores to the FEED terminals of the

first switch and link its LOAD terminals to the FEED terminals of the second switch.

4 Complete the wiring back to the consumer unit as for Option 1.

Wiring Regulations

Calculation of likely current demand and of cable size is quite complicated. You do not need to provide a circuit for the theoretical maximum power usage of the cooker - perhaps as much as 60 amps - because you are unlikely to ever use all the hobs, the oven and grill at full power simultaneously. A principle called diversity is used to calculate a realistic current demand: 10 amps plus 30 per cent of the remainder of the total theoretical maximum demand (plus 5 amps if you are using a cooker control

unit with a socket outlet).

As a general guide, for cookers rated at up to 12kW you should use 6mm² cable with a 30/32-amp protective device and an isolating switch of the same rating. Use 10mm² cable if the circuit is over 20m (66ft) long, up to a maximum of 30m (98ft). For cookers rated at over 12kW, use 10mm² cable, a 40/45-amp protective device and a 45/50-amp switch. Use 16mm² if the circuit is over 6.7m (22ft) long, up to a maximum of 10.2m (33ft).

WIRING SHOWERS

The instantaneous electric shower has proved a popular alternative to the conventional shower since it is much easier to install, needing just one mains plumbing connection plus a simple-to-run circuit from the consumer unit. Like a cooker, it must have its own circuit since the current demand is high.

THINGS YOU NEED

- Double-pole switch to suit shower wattage
- 6mm² two-core-and-earth cable
- Green/yellow PVC earth sleeving
- 4mm² single-core earth cable
- Earth clamp
- RCD and en-closure
- Electrical tools
- General d-i-y tools plus tools and materials for plumbing in water supply

The circuit is quite straightfor-ward; the cable runs from the consumer unit to a double-pole isolating switch and then on to the shower unit itself, ideally passing through the wall immediately behind the unit. The switch is usually a ceiling-mounted cord-operated type, situated within the bath or shower room, and should have a neon indicator light. It must also have a mechanical on-off flag to show whether the power is on or off if the neon light fails. The circuit should have RCD protection.

This can be done by running the circuit cable via an in-line RCD mounted in its own

Wiring Regulations

Use 6mm² cable to wire most shower units. This can supply units rated at up to only 8kW if the protective device is a 45-amp rewirable fuse; this is not recommended anyway if the circuit is being newly installed. However, it can supply up to 10kW if a 45-amp cartridge fuse or MCB is used – enough to cope with any future developments in more powerful heaters. You need to use 10mm² cable if the circuit is over 20m (66ft) long.

Use an RCD to protect the circuit; otherwise the type of earth-ing system your house has may further limit the maximum cable lengths allowed for any particular size of cable.

A wall-mounted switch can be used instead so long as it is at least 2.5m (8ft 3in) away from both the shower and the bath (it can be within 600mm – 2ft – of the closed sides of a shower cubicle in a room without a bath), or else is situated outside the room where the shower is fitted. Using one can make for shorter (and therefore cheaper) cable runs than are required for a ceil-ing-mounted switch, and can make the wiring work easier too because access is not needed to the ceiling void.

The rating of the switch used and the circuit protective device fitted depend on the shower's wattage. For units rated at up to 7kW, use a 30/32-amp switch and protective device, and a 40/45-amp switch and protective device for more powerful units.

If the house already has a circuit to an electric cooker and the main service fuse is rated at only 60 amps, get expert advice in case adding a shower circuit as well leads to overloading of the house's incoming supply cable.

Lastly, it is essential that the supply pipework to the shower unit is cross-bonded to earth. This means fitting an earth clamp on the pipe and running a 4mm² single-core earth cable back to the main earthing connection at the consumer unit.

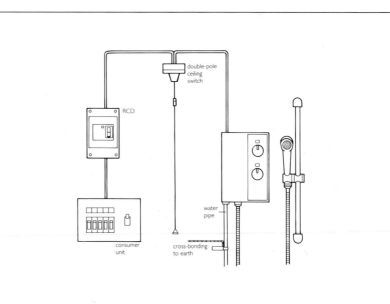

enclosure; alternatively, if there is no spare fuseway or MCB available, the shower circuit can be protected by a suitably-rated RCBO – a combined MCB and RCD – again in a separate enclosure. The RCD/RCBO should be the high-sensitivity type with a 30mA trip current rating.

1 Arrange the cable route to the shower unit so that it emerges through the wall immediately behind where the unit will be mounted. Fit the unit and complete the plumbing first, then feed the cable into it via any seal fitted to the casing. Connect the cores to their respective terminals at the heater terminal block, and secure the sheathing in the cable grip if one is fitted.

2 Run the cable to the switch position, and connect its cores to the LOAD and earth terminals. If the switch is ceiling-mounted, check that it is securely fixed either to a ceiling joist or to a batten fixed between adjacent joists.

3 Run the cable back from the FEED and earth terminals of the switch to the consumer unit, via the RCD. Turn off the power at the consumer unit, connect its live core to the fuseway or MCB, and take the neutral and earth cores to their respective terminal blocks.

4 Fit an earth clamp to the shower supply pipework and connect the 4mm² single-core earth cable to it. Run the cable back to the consumer unit and connect it to the main earthing terminal.

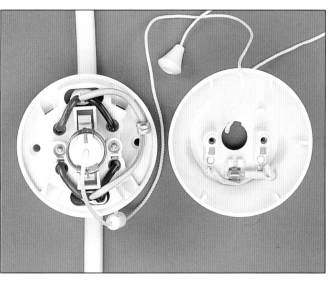

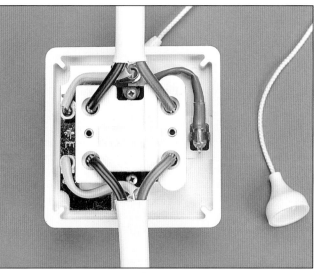

Above and left: double-pole ceiling switches

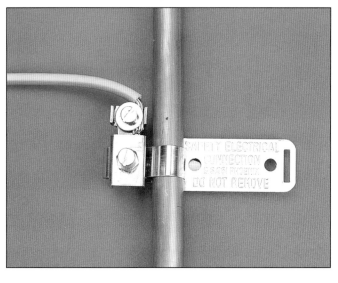

Left: Earth clamp on shower supply pipework

WIRING BELLS AND ALARMS

Although battery-powered door bells, burglar alarms and smoke detectors are still popular, there is a growing trend towards supplying all three from the mains. The chief advantage of using mains power is greater reliability.

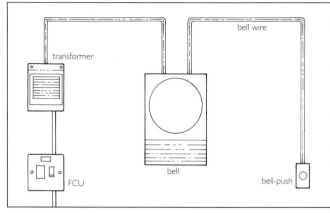

THINGS YOU NEED

- Bell and bell-push
- Bell wire
- 1mm² two-core-and-earth cable
- Green/yellow PVC earth sleeving
- Junction box or fused connection unit for mains supply connection
  or
- Spare 5/6-amp fuseway or MCB
- Sleeving for bell wire
- Electrical tools
- General d-i-y tools

Installing mains-operated door .ells and smoke detectors could not be simpler, and although some early battery-powered burglar alarm kits did involve a lot of fiddly wiring to the door and window contacts, their modern counterparts are far less complicated. This is mainly due to the general changeover from window contacts to PIR sensors which detect movement inside the house. There is very little additional wiring involved in providing a mains supply for any of these pieces of equipment, which you can supply either from their own fuseways at the consumer unit or, in the case of door bells and burglar alarms, from existing circuits.

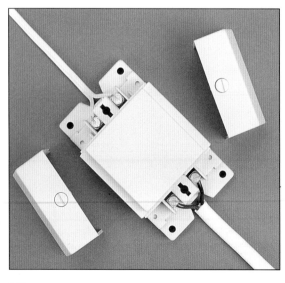

Left: Bell transformer

DOOR BELLS

A mains-powered door bell actually operates at extra-low-voltage, taking its power supply from a suitable transformer. If there is no spare fuseway or MCB available for use as a separate bell circuit, the simplest solution is to supply the transformer from the most conveniently sited lighting or power circuit.

I Install the bell-push and the bell baseplate, linking the two with bell wire connected to the terminals as detailed in the manufacturer's instructions. Then run bell wire back to the transformer and connect it to the appropriate secondary terminals.

2 Run 1mm² two-core-and-earth cable from the primary terminals of the transformer to your chosen main circuit connection point. This can be at an existing loop-in ceiling rose or main junction box if there is room to connect in an extra cable, at a new junction box cut into the lighting circuit wiring (see page 80), or at a fused connection unit (FCU) wired up as a spur from a power circuit (see

pages 96-7). If the last option is used, fit a 3-amp fuse in the FCU. Turn off the power at the consumer unit and check that the circuit is dead, then make the final circuit connection.

3 If you have a spare 5/6-amp fuseway or MCB available in the consumer unit turn off the power at the consumer unit and wire the live core of the cable from the bell transformer directly into it Take the neutral and earth cores to their respective terminal blocks.

4 If you are increasing your consumer unit capacity anyway (see page 119), you can install a bell transformer within the new unit, supplying it from a 5/6-amp fuseway or MCB using a short loop of 1mm² cable within the consumer unit. The live cable core links the MCB to one primary terminal of the transformer while the neutral and earth cores link the other transformer terminals to the neutral and earth terminal blocks within the consumer unit. Note that the outgoing bell wire connected to the secondary terminals must be sleeved or segregated within the consumer unit, and must not be run to the bell in the same conduit as any mains-voltage cables.

BURGLAR ALARMS

Most modern burglar alarm systems run on mains-voltage electricity, and contain a rechargeable battery to provide back-up in the event of a power failure. The power supply can be provided in exactly the same way as for a mains-powered door bell (see opposite). If an MCB is used, it should not have RCD protection and should be fitted with a locking device to prevent the MCB from being accidentally switched off, so disarming the system.

1 Install the kit as detailed in the manufacturer's instructions, then connect the mains supply cable into the system's control unit.

2 Run the supply cable back to your chosen connection point, either to an existing circuit or at the consumer unit. Turn off the power at the consumer unit and check that the circuit is dead. Make the final connections as for door bells (see Steps 2 and 3 opposite).

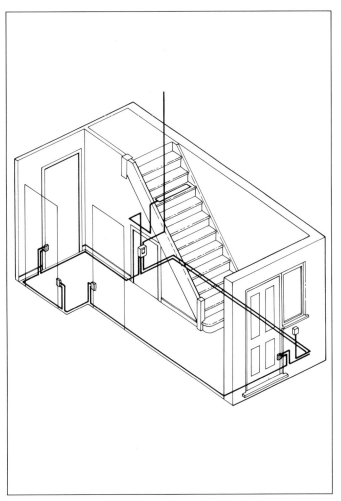

THINGS YOU NEED

- Burglar alarm kit
- 1mm² two-core-and-earth cable
- Green/yellow PVC earth sleeving
- Junction box or fused connection unit for final circuit connection

or

- Spare 5/6-amp fuseway or MCB
- Locking device for MCB
- Electrical tools
- General d-i-y tools

Left: Typical burglar alarm circuit with door and window sensors and pressure pads linked to the control unit

SMOKE DETECTORS

In 1991 The Smoke Detectors Act made it mandatory for all new or refurbished properties to have a smoke detector fitted on each floor level. The 1992 edition of the Building Regulations now requires them to be mains-operated. If you do not have battery-operated smoke detectors in your home, have only one detector, or have more but do not have them interconnected, it is well worth changing over to a mains-operated system using detectors with battery back-up complying with British Standard BS5446. These are interconnected and wired back to a 5/6-amp fuseway or MCB in the consumer unit. As with burglar alarms, the circuit should not be RCD-protected.

1 Install and interconnect the detectors as detailed in the manufacturer's instructions, and run the circuit cable back to the consumer unit.

2 Turn off the power at the consumer unit and link the live cable core to the fuseway or MCB. Take the neutral and earth cores to their respective terminal blocks.

THINGS YOU NEED

- Smoke detectors
- Cable as recommended by detector manufacturer
- Green/yellow PVC earth sleeving
- Spare 5/6-amp fuseway or MCB
- Electrical tools
- General d-i-y tools

WIRING STORAGE HEATERS

Electric space heating is most economically supplied by storage heaters, which can be charged at night or during the early morning with low-priced off-peak electricity and then discharge the stored heat during the day. Each heater is wired on its own radial circuit.

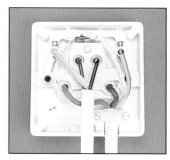

THINGS YOU NEED

- Off-peak consumer unit with 20-amp or 30/32-amp fuseways/MCBs as appropriate
- Storage heater(s)
- 20-amp or 30/32-amp double-pole switch(es) as appropriate
- Two-core-and-earth circuit cable sized to match heater wattage
- Three-core heat-resistant flex
- Switched fused connection units fused as appropriate
- 2.5mm² two-core-and-earth cable for wiring spurs to FCUs
- Electrical tools
- General d-i-y tools

Top: Double-pole switch with flex outlet

Above right: Storage heater circuit (left) and wiring for storage heater with a fan powered by the 24-hour supply (right)

The circuits to storage heaters are wired from a separate off-peak consumer unit. For heaters rated at up to 4kW, the circuit is ·wired in 2.5mm² cable from a 20-amp fuse or MCB. It terminates close to the heater position at a 20-amp double-pole switch with flex outlet, from where a short length of 2.5mm² three-core heat-resistant flex runs to the heater itself. For heaters rated at over 4kW, 4 or 6mm² cable is used and the circuit is run from a 30/32-amp fuse or MCB to a switch of the same rating.

Fan-assisted storage heaters also need a continuous power supply for the fan, and this is usually provided by a switched fused connection unit (FCU) wired as a spur from a nearby circuit supplying socket outlets. The FCU is fitted with a 5-amp fuse, and heat-resistant flex links it to the fan terminals in the heater.

An alternative way of wiring fan-assisted heaters is via a special twin double-pole switch. This is rated at 25 amps and has two flex outlets, one for the heater and one for the fan; the single on-off switch isolates both components when required. The fan is again supplied via a fused spur.

Combined storage/convector types also need a 24-hour power supply so that the convector can provide additional heating during the day if the stored heat is exhausted. The supply is usually provided by a circuit wired direct from the consumer unit. Again, heat-resistant flex is used for the final connection to the heater.

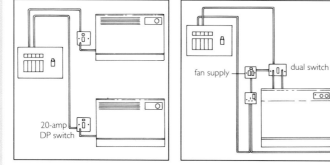

20-amp DP switch

fan supply   dual switch

I Install the heater in the desired position, and fit its double-pole control switch next to it. Link the heater's terminal block to the LOAD terminals of the switch with heat-resistant flex.

2 Run the circuit cable back from the FEED terminals of the switch to the off-peak consumer unit. Turn off the power at the consumer unit and connect the live cable core to an appropriately rated fuseway or MCB. Take the neutral and earth cores to their respective terminal blocks.

3 If the heater is fan-assisted, provide its 24-hour power supply by installing a spur from a nearby socket outlet circuit. Run the 2.5mm² spur cable to an FCU fitted next to the heater switch, and link FCU and fan terminals with heat-resistant flex. Fit a 5-amp fuse in the FCU.

If a 25-amp twin switch is used to control both fan and heater, take the circuit cable to the heater side of the switch, and run 1mm² cable from the FCU to the fan side of the switch. Use heat-resistant flex as before to connect the heater and fan to the switch.

4 If the heater is the combined storage/convector type, run a circuit from a spare 20-amp fuseway in the 24-hour consumer unit for the convector heater.

Off-peak tariffs

Electricity companies are developing new ways to supply off-peak electricity. This means that through the years up to 1997 or 1998 the current off-peak supplies such as Economy 7 may be phased out, certainly as far as new installations are concerned. Check with your local electricity supplier for more information before installing any circuits or equipment intended for use with off-peak electricity.

So long as your home has a master socket outlet (called a linebox) connected to the incoming line by your telecommunications operator, you can extend the system within the house yourself in one of two ways. The first is to wire special telephone cable into the linebox, and the second is to plug a converter into it. The converter still allows a telephone to be plugged in at the linebox position, but does not look as neat as a wired-in extension cable.

From the linebox, the cable is run to the new socket outlet positions. Since it is so thin, it can generally be hidden easily below the bottom edge of skirting boards and round door architraves; it must not, however, be run in the same conduit as any mains-voltage cables and should be kept at least 50mm (2in) away from them on its way from outlet to outlet. The circuit should not exceed 50m (160ft) in length, measured from the linebox to the most remote outlet.

You can run the cable to the outlets in series, looping it from outlet to outlet, or in parallel, using a junction box which will accept up to three extension cables. Use whichever makes the most economical use of cable.

Lastly, if you want to connect additional equipment such as a fax machine or a computer modem to individual outlets, you can either use a plug-in two-way adaptor or fit a twin outlet.

If you have two incoming lines, they will terminate at a dual master socket; make sure in this situation that you know which line you are extending. Outlets come in finishes to match most wiring accessory ranges, and can be flush- or surface-mounted.

1 Decide where the extension outlets will be sited, and install a mounting box at each position. Run the cable from the master socket to each extension position, clipping it at 300mm (12in) intervals and leaving about 75mm (3in) at each outlet for ease of connection.

2 Slit the sheathing lengthways with a knife, peel it back by about 50mm (2in) and cut off the waste. Inside are six cores colour-coded as follows:

1 green with white rings
2 blue with white ring
3 orange with white rings
4 white with orange rings
5 white with blue rings
6 white with green rings

Cores 1 and 6 are not needed on simple domestic installations and may not be included in the cable sold for this purpose. But connect them if they are included.

Connect each one to its matching numbered terminal on the extension outlet faceplate, using the connector tool to push the cores into the terminals. Do not try to strip the core insulation; the jaws inside the terminal strip it to make the electrical contact as you push them in. Fold the cores neatly into the mounting box and attach the faceplate.

WIRING TELEPHONE POINTS

The introduction of plug-in telephone socket outlets means you can have one in every room in the house (although installation in bathrooms is not recommended). However, you should not plug more than four telephones in at once, or none of them will ring properly.

3 To extend from this outlet, connect in the cores of the second cable in the same way, linking like cores to like cores.

4 To use a junction box, connect in the supply cable first. Note that all six cores go to the numbered in-line terminals down one side of the box. Connect in the first extension cable to the same six terminals, connecting like to like. Then connect the third (and, if needed, the fourth) cable in the same way to the opposite terminal bank.

5 Make the final connection at the master socket outlet, either by plugging in the connector or by removing the faceplate and wiring the extension cable cores into its terminals, as always linking like cores to like cores.

THINGS YOU NEED

- Secondary telephone socket outlets
- Flush or surface mounting boxes
- Extension cable
- Plug-in converter for master outlet
- Cable clips
- Handyman's knife
- Connector tool
- General d-i-y tools

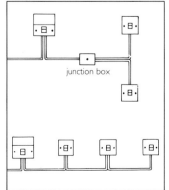

junction box

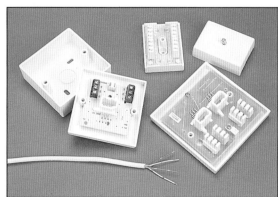

WIRING TV/FM AERIAL SOCKETS

TV sets are likely to be found in any room except the bathroom. Unless you live in an area with good reception, each will need connecting to a loft or rooftop aerial at a socket outlet, as will an FM radio.

THINGS YOU NEED

- Low-loss UHF coaxial cable
- Coaxial 'male' and 'female' plugs
- Coaxial socket outlets
- Signal splitter for two-outlet system
- Amplifier for more than two outlets
- Handyman's knife
- General d-i-y tools

Below: TV and FM signals run via a single downlead when diplexers are used

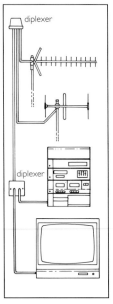
diplexer

diplexer

The basic aerial/socket outlet arrangement in a one-outlet home has a coaxial cable running down inside or outside the house from the aerial to the outlet position, where it is then connected into the socket outlet. A short plug-in flylead then links the TV or FM radio to the outlet. Double outlets are used to allow the TV and an FM radio to be plugged into the same outlet, which is wired with two downleads, one from each aerial. Alternatively, special double outlets containing a diplexer can be used; one is installed next to the aerial, the other at the TV/radio position, and the two outlets are linked with a single coaxial downlead which carries both signals.

If you want more than one aerial socket outlet you need a device called a signal splitter, and possibly a signal booster or amplifier too if the incom-

ing signal is weak. A splitter divides the aerial download in two, and while in theory you could use further splitters on each sub-lead the signal strength will rapidly diminish. If you want more than two TV outlets, it is better to connect the aerial to a mains-powered amplifier with as many outlets as you require and then to run a separate download to each outlet position. Amplifiers are available in different versions to supply just TV signals to each outlet, or to distribute both TV and FM signals from their respective aerials to double outlets, each incorporating a diplexer.

An extra bonus with a TV distribution system of this type is that video playback can also be provided to each TV set. To do this, the TV aerial downlead runs first to the video recorder's input socket as on an unamplified system. The output socket is then connected to the amplifier, and leads are run from there to the various outlet positions. Each TV set is then tuned to the video channel.

1 Install the TV and/or FM aerials, then work out how many outlets you want and whether each outlet position is to have both TV and FM signals or just TV. Buy the appropriate components.

2 Install the outlets, which may be surface-mounted or fitted over a standard 25mm-deep flush mounting box, and run coaxial cable to each one from the aerial, the amplifier or the video recorder, depending on what distribution system you have selected. Connect the cables as appropriate to the various components. Finally, connect the amplifier (if fitted) to its power supply; it can be plugged into a socket outlet if it is inside the house, but should be wired via a fused connection unit on a spur from a socket outlet circuit if it is in the loft.

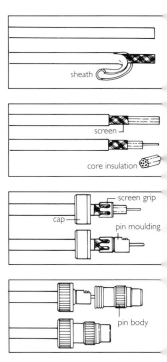

sheath

screen

core insulation

screen grip

cap

pin moulding

pin body

Above: To prepare coaxial cable slit the sheath, peel it back and remove 50mm (2in). Roll back the screening wires for 25mm (1in) over the pin insulation, then cut off 12mm (1/2in) of this to reveal the pin. Fit the cap, screen grip and pin moulding, then slip on the pin body and screw on the cap

Below: Standard double outlet

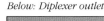

Below: Diplexer outlet

INSTALLING NEW CIRCUITS

It is often possible to alter or extend your existing circuits to provide the lighting or power supply you need. However, if the lighting circuits are fully loaded or socket circuits are already covering the maximum permitted floor area, you will need to add new circuits.

If your existing consumer unit has no spare fuseways available for the new circuits you can either add a small unit or replace the original unit with a larger one. This gives you the opportunity to install modern equipment and to plan for future expansion.

Start by deciding on where the new lights are to go. Then relate their positions to the existing circuit wiring; in certain circumstances it may make more economical use of cable to disconnect some existing lights and to supply them from the new circuit so that the existing circuit can itself be extended to feed some of the new light positions.

Next, plan the circuit cable route carefully to minimise the disruption caused by having to lift floorboards.

Do not bunch new and existing circuit cables where they pass through obstructions such as joists; drill a fresh hole for each new cable. Similarly, do not try to feed additional cables into existing conduit on switch drops; it is better to cut new chases down to new switch positions.

The circuit itself is run in 1mm² two-core-and-earth cable from a spare 5/6-amp fuseway or MCB in the consumer unit. If you do not have a spare way, install a new consumer unit (see page 119) to supply the new circuit. Remember that a qualified electrician must make the final connection to the new unit.

1 Start by installing the new ceiling roses and light fittings and their switches. Fit any junction boxes that are required along the planned cable route.

2 Start wiring the circuit from the most remote light, using loop-in or junction-box wiring as appropriate. Cover all bare earth cores with green/yellow PVC sleeving and use red tape as a 'live' indicator on all black, blue and yellow switch cores. If switches are being fitted on metal mounting boxes, connect the faceplate and box earth terminals with an earth link.

ADDING A NEW LIGHTING CIRCUIT

If you want to install extra lights and your existing lighting circuits are already supplying their practical maximum of eight lighting points, you will have to provide an additional circuit wired from the consumer unit or a separate new unit.

3 At the consumer unit, turn off the power and connect in the circuit cable cores. The live core goes to the fuseway on the circuit fuse or MCB, while the neutral and earth cores go to their respective terminals. Sleeve the earth core.

From left: Use loop-in or junction box wiring. Make connections as shown (other circuits omitted for clarity)

THINGS YOU NEED

- 1mm² two-core-and-earth cable
- Three-core-and-earth cable for two-way switching
- Green/yellow PVC earth sleeving
- Red PVC insulating tape
- Wiring accessories as required
- Spare 5/6-amp fuseway/MCB or
- Small consumer unit and splitter box and 25mm² meter tails, plus 16mm² single-core earth cable
- Electrical tools
- General d-i-y tools

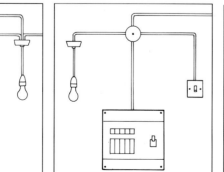

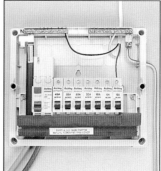

ADDING SOCKET OUTLETS FOR LIGHTING

Fixed wall and ceiling lights have their own circuits, but table and standard lamps generally have to share socket outlet circuits with other electrical appliances. Where socket outlets are in short supply an alternative is to add a new sub-circuit supplying just the plug-in lights.

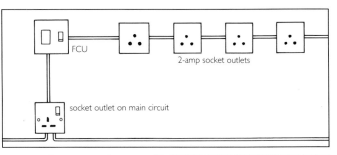

Below: Wire the socket outlets as a radial circuit supplied via an FCU. The photograph shows a 2-amp round-pin socket outlet

FCU

2-amp socket outlets

socket outlet on main circuit

THINGS YOU NEED

- 2-amp unswitched socket outlets
- Mounting boxes
- 1mm² two-core-and-earth cable
- Green/yellow PVC earth sleeving
- Switched fused connection unit fitted with 3-amp fuse
- 2.5mm² two-core-and-earth cable for spur to FCU
- 2-amp round-pin plugs
- Electrical tools
- General d-i-y tools

Table and standard lamps take relatively little current, so do not need to be supplied by a 13-amp socket outlet. They are generally used in living rooms where the demand for socket outlets is often high. A neat solution is to supply all the plug-in lamps from their own dedicated socket outlets, which are wired up as a sub-circuit from the existing socket outlet circuit. This arrangement has the added advantage of allowing you to control all the lamps from just one switch position instead of having to turn them all on and off individually.

The socket outlets used are unswitched, rated at 2 amps, and accept small round-pin plugs like those used to connect small appliances to old-style 2-amp power circuits before the advent of the ring circuit. Modern types are designed to fit on a standard single mounting box.

Each lamp is permanently plugged into its own socket outlet sited nearby to keep flex lengths to a minimum, and the sub-circuit is wired as a radial circuit running from a switched fused connection unit (FCU) to each socket outlet in turn. The FCU is supplied as a spur and acts as the on-off switch for all the lamps, although individual lamps can still be turned on or off at their own switches, and it is fitted with a 3-amp fuse. This means that the sub-circuit can supply up to 700 watts – more than enough for the number of lamps likely to be used.

This type of circuit is also ideal for other low current users which need a lot of socket outlets, such as hi-fi systems.

1 Decide where the new 2-amp socket outlets are to be fitted, and install a flush or surface mounting box at each position. Fit a mounting box for the FCU at a point close to where the spur will be connected into the existing socket outlet circuit.

2 Run 1mm² cable from the FCU position to each outlet position in turn, looping it into and out of each box until the most remote outlet is reached. The cable can be run in mini-trunking at skirting-board level, or beneath the floorboards if lifting one or more of them does not cause major disruption in the room concerned.

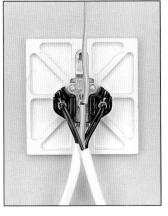

3 Connect the cables to each 2-amp socket outlet in turn, and fit each one to its mounting box. At the FCU, connect the sub-circuit cable to the LOAD terminals, and run a length of 2.5mm² two-core-and-earth cable from its FEED terminals back to the chosen connection point on a nearby circuit. Fit a 3-amp fuse in the FCU.

4 Turn off the power at the consumer unit and check that the circuit is dead. Connect in the spur at a 13-amp socket outlet or via a junction box cut into the circuit cable (see pages 92-3 and 96-7 for more details), whichever is more convenient.

If extending your home and its wiring circuits means that a socket outlet circuit would be supplying rooms with a floor area greater than that allowed by the Wiring Regulations (see below), you must install a new circuit to serve the additional floor area. You may also decide that a separate circuit to kitchen and utility room socket outlets is worth having.

As with adding a new lighting circuit, you need either a spare fuseway in your consumer unit or a new unit installed alongside it to supply the new circuit. What type of circuit you install will depend on the distance from the consumer unit to the room or rooms being served; running a ring circuit to the far side of the house and back again could be expensive, so it is worth checking whether a 20-amp radial circuit can supply the necessary demand. However a ring circuit is cheaper than a 30-amp radial circuit run in 4mm² cable.

Start by working out where the new socket outlets are required, then check the floor area the new circuit will be serving. Remember that a ring circuit can supply an area of 100sq m (1075sq ft), while a 20-amp radial circuit can sup-

ply only 20sq m (215sq ft). Next, plan the cable route from the outlets back to the consumer unit, with spurs added to the circuit where appropriate to save on cable. If you do not have a spare fuseway for the new circuit, you will again need a new unit to supply it. You need a 30/32-amp fuse or MCB for a ring circuit, and a 20-amp one for a radial circuit wired in 2.5mm² cable.

1 After deciding where the new socket outlets and fused connection units (FCUs) are required, fit a mounting box at each position and run the cable from box to box, leaving a generous cable loop for connection to each accessory.

2 Add spurs to the circuit where this avoids unnecessary use of cable. As you are installing a new circuit, it is better to connect these in using junction boxes rather than at socket outlets; these then remain free for any future extensions to the circuit you might want to make.

3 Strip the cables and connect the cores to the relevant terminals on each acces-

ADDING A NEW SOCKET OUTLET CIRCUIT

Since modern socket outlet circuits can supply an unlimited number of outlets within a given floor area, you should be able to meet your needs by adding extra outlets to existing circuits. However, if you extend your home you may have to provide an additional circuit to supply socket outlets within the extension.

sory faceplate. Sleeve the earth cores first. If the faceplate is mounted on a metal box, connect the faceplate earth terminal to the box terminal with a short earth link.

4 Run the circuit cable(s) back to the consumer unit if you have a spare fuseway, and turn off the power. For a ring circuit, fit a 30/32-amp fuse or MCB and connect both cable live cores to the fuse/MCB terminal. For a radial circuit wired in 2.5mm² cable, fit a 20-amp fuse/MCB and connect the live core to it. In each case, connect the neutral and earth cores to their respective terminal blocks. Sleeve the earth core(s) first.

THINGS YOU NEED

- 2.5mm² two-core-and-earth cable
- Green/yellow PVC earth sleeving
- Socket outlets
- Fused connection units
- Mounting boxes
- 30-amp junction boxes
- Spare fuseway or
- New consumer unit, splitter box and 25mm² meter tails plus 16mm² single-core earth cable
- Electrical tools
- General d-i-y tools

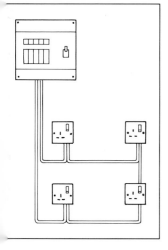
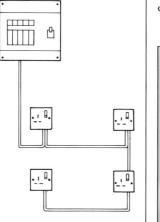

From left: wire the new circuit as a ring or radial circuit as appropriate. Make the connections as shown for a ring circuit (other circuits omitted for clarity)

ADDING RCD PROTECTION

The use of residual current devices (RCDs) to protect electrical installations and their users has grown dramatically in recent years, and all newly wired homes will now incorporate RCD protection for some or all of the circuits. Many older systems do not, however, and they would be much safer if such protection were added.

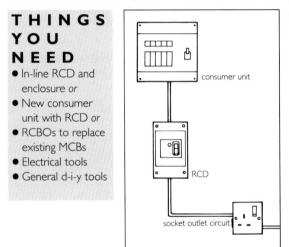

consumer unit

RCD

socket outlet circuit

RCDs (formerly known as current-operated earth leakage circuit breakers or ELCBs) protect wiring systems and their users by detecting earth fault currents smaller than those required to cause overcurrent protection devices such as fuses or MCBs to operate. They do not prevent you from receiving a shock through direct or indirect contact with live parts, but operate quickly enough to prevent the shock from being fatal. It is therefore well worth considering fitting one to an unprotected system, or at least to the most at-risk parts of it.

Early voltage-operated circuit breakers were designed only to give protection against indirect contact with live parts (for example when a metal casing on an appliance becomes live), and are now no longer fitted to domestic wiring installations. If you have such a device, it should be replaced by a modern RCD.

You can add an RCD to your wiring system in a number of ways. The whole system can be protected either by fitting

one between the electricity meter and the consumer unit or by installing a consumer unit with the RCD as the main isolator. This is not now recommended because a fault on one circuit will cut the supply to the whole house, which contravenes the latest Wiring Regulations. You can protect individual circuits in a similar way, either by positioning the RCD on the outgoing circuit cable at the consumer unit or by fitting a combined RCD and MCB (known as an RCBO) in the consumer unit as the protective device for the circuit concerned. This last option is open to you only if you are replacing your consumer unit or if your existing unit has MCBs and the consumer unit manufacturer makes compatible RCBOs.

With new wiring installations and rewires, the favoured procedure is to fit a split-load consumer unit, with a main isolator switch controlling the whole system and the RCD protecting only the circuits supplying socket outlets, cookers and electric showers. See page 120.

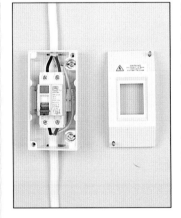

I To add an in-line RCD to an individual circuit, turn off the power at the consumer unit and disconnect the circuit

cable from it. Mount the RCD enclosure next to the consumer unit and connect the circuit cable to its lower (outlet) terminals and to the earth

terminal. Then add a short length of cable of the correct size for the circuit between the upper (input) and earth terminals of the RCD and the terminals on the consumer unit.

2 If you are installing a new consumer unit, fit the RCD to protect the circuits supplying socket outlets, a cooker or an electric shower. See pages 119 and 120.

3 If you want to add whole house protection to an existing installation by installing an RCD on the incoming supply, you must leave the job to an electrician.

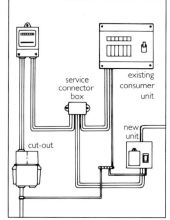

FITTING AN EXTRA CONSUMER UNIT

If you want to add extra circuits to your wiring system, but you have no spare fuseways in your consumer unit, the most economical method is to add a small consumer unit to provide the additional fuseways you need. Units are available fitted with fuses or MCBs.

If you need to install an extra consumer unit, it makes sense to fit one that can cope with any future expansion of the system as well as supplying the extra fuseways you need now. If you are adding circuits to high-powered appliances such as cookers and showers, check first with your electricity supplier to ensure that there is no risk of overloading your supply.

The power supply to the new unit comes via a pair of new 25mm² meter tails and a special junction box called a **service connector box** (also known as a splitter box or Henley box). You can fit the box and the meter tails between it and the new consumer unit, but you must call in a qualified electrician to disconnect the meter tails supplying your existing consumer unit, reconnect them to the connector box and then add new tails from the box to the existing consumer unit.

The electrician will have to obtain permission from the electricity supply company to break the seal on the service fuse and remove it before starting the work. The electricity company will have to reseal the fuse, for which a charge will be made.

1 Mount the new unit close to the existing consumer unit or fusebox, on a fireproof mounting board.

2 Connect the live cores of the cables supplying the new circuits to their respective fuseways or MCBs in the new unit and take the circuit neutral and earth cores to their respective terminals.

3 Mount the connector box on the new backing board where the existing meter tails will be able to reach it easily and connect it to the supply terminals of the new unit with two new 25mm² meter tails, the live one colour-coded red, the neutral one black.

4 Link the new unit's earth terminal to the house's main earthing point with a length of 16mm² sheathed earth cable.

5 Call in a qualified electrician to finish the work.

Top left: Two-way consumer unit. Right from top: One-way consumer unit; connector box – this comes in two parts, each containing a bank of terminals for the live or neutral meter tails

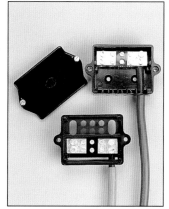

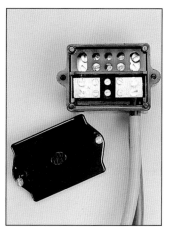

THINGS YOU NEED

- Small consumer unit with appropriate fuses or MCBs and RCD if required
- Qualified electrician
- Fireproof mounting board
- Connector box
- 25mm² meter tails
- 16mm² single-core earth cable
- Electrical tools
- General d-i-y tools

Safety Warning

Note that only one circuit is allowed to be connected to each fuseway in a consumer unit. Never try to connect more circuits to a consumer unit than it is designed to accept.

REPLACING A CONSUMER UNIT

If you are carrying out extensive additions to your house wiring, it makes sense to fit a completely new consumer unit rather than add a small one to supply the extra circuits, especially if you currently have an old-fashioned fusebox with rewirable fuses. Contact your electricity supply company to check that there is no risk of overloading the supply with the additional circuits.

THINGS YOU NEED

- New consumer unit with MCBs and RCDs as required
- Qualified electrician
- Non-mains lighting
- Electrical tools
- General d-i-y tools

The modular design of modern consumer units offers total flexibility, allowing you to provide individual circuits to suit your electrical requirements. For example, apart from the traditional lighting, socket outlet, immersion heater and cooker circuits you can have extra circuits supplying your freezer, your burglar alarm or security lighting system, your garden workshop, your doorbell, even your smoke alarms. You can also select which parts of the system to protect against the risk of earth fault currents, and make provision for any foreseeable future expansion of the system.

The use of an RCD as the main isolator in the consumer unit ensures that all circuits have earth fault protection. However, such an arrangement suffers from the drawback that a fault on one circuit shuts down the entire system, which contravenes the latest Wiring Regulations. For this reason current practice is to install split-load units, with high-sensitivity RCD protection provided only for circuits that really need it – those supplying socket outlets, cookers and showers inside the house and any electrical equipment installed in detached outbuildings. Circuits supplying lighting, burglar alarms, bells, smoke detectors and even domestic freezers are either not RCD-protected, or have a lower-sensitivity 100mA RCD to give protection against indirect contact (contact with parts made live by an electrical fault) as opposed to direct contact with the live parts of the wiring system. This lower-sensitivity RCD also has a time-delay mechanism to prevent it from tripping when the high-sensitivity device is activated.

Once you have selected the components you require, the changeover involves the relatively straightforward but fairly time-consuming task of disconnecting everything from the old consumer unit or fusebox and connecting all the existing circuits (and any new ones you are adding) into the new unit. Unless the cable feeds to the existing unit have been left unusually long, you will have to remove the old unit before you can start making connections to the new one; this will mean calling in an electrician to disconnect and make safe the meter tails, and to reconnect them again when you have completed the changeover. The electrician will need the permission of the electricity supply company to remove the seal on the service fuse and remove it before carrying out the work. The electricity supply company will have to reseal the fuse when the work is completed, for which you will be charged.

1 Have the power supply and earth connections to the existing consumer unit disconnected and made safe by a professional electrician. Arrange adequate battery or bottled-gas lighting to enable you to work safely without mains lighting. Then systematically label and disconnect each circuit cable in turn and remove the old consumer unit or fusebox.

2 Mount the new unit on the backing board. Connect each circuit cable to the new unit in turn, working along the live busbar from the main isolator. Connect the neutral and earth cores to their respective terminal blocks first, then link the live core to the MCB and clip it into place on the busbar. Double-check the soundness of each connection as you work. Then fit the busbar shield, and blank off any unused fuseways.

3 Replace the earth connection between the unit and the system's main earthing point. Check that there are equipotential bonding conductors from gas and water pipework connected to the earthing point; install them if not (see page 66). Then have the power reconnected by a professional electrician, fit the unit's cover and check that all circuits have power. Test the operation of any RCDs fitted, then fill in the identity of each MCB on the label inside the consumer unit.

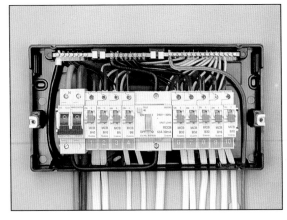

WIRING OUT OF DOORS

Electricity can be just as useful in the garden as it is in the house. It can light the way to your front door, enable you to use the patio for evening entertainment when the weather is fine, show off your plants and even help to deter would-be intruders. It can also drive the growing range of garden power tools now available, and can supply greenhouses and garden workshops. However, the potential danger outdoors is greater than inside the house, so all wiring work must be carried out to the highest standards and a full range of protective devices must be used to ensure the highest levels of system and user safety.

There are several outdoor electrical projects that are well worth carrying out. The simplest is to install a light by your front door, so visitors can see their way after dark and you can identify them easily. The same wiring techniques can be used to add lights elsewhere on the house walls, to light up side passages, a patio, the approach to a garage and so on. The fittings used can be decorative or functional, and for your convenience and security they can be automatically controlled in various ways.

Lights in these situations can be supplied as extensions of RCD-protected indoor circuits, but if you want lights remote from the house – to light the drive or path, or for illumination down the garden – they must be wired up on their own circuit unless they are the extra-low-voltage type run from a transformer.

The same rules apply if you want to install socket outlets out of doors to supply garden power tools or appliances such as kettles and toasters used for al fresco meals. An outlet on the house wall can be supplied as a spur from an indoor socket outlet circuit, although it must be RCD-protected. Outlets remote from the house must have their own 30/32-amp circuit, again protected by a high-sensitivity RCD.

Lastly, a power supply to detached outbuildings such as greenhouses, sheds and workshops can add greatly to their usefulness. You can have light so you can work after dark, heat for extra comfort in cold weather both for you and your plants, and the convenience of being able to use power tools. Once again, the supply must be via a separate circuit run from the consumer unit and protected by an RCD. It can be taken to the building overhead or underground, whichever is the more convenient. The circuit must be earthed independently via an earth rod.

FITTING LIGHTS ON THE HOUSE WALL

The simplest way to install outside lights is to mount them on the house wall. The main advantage of this is that you can extend your house wiring to supply them; lights remote from the house have to be run on a separate circuit (see pages 128-9) unless they are the extra-low-voltage type (see page 131).

THINGS YOU NEED

- New light fitting
- Round conduit box (some fittings only)
- 16mm round PVC conduit
- Strip connectors
- 1mm² two-core-and-earth cable
- Green/yellow PVC earth sleeving
- Switched fused connection unit fitted with a 5-amp fuse (Option 1)
- 2.5mm² two-core-and-earth cable for spur to socket outlet circuit (Option 1)
- Four-terminal junction box
- One-way or two-way switch(es)
- Red PVC insulating tape
- Silicone mastic
- Electrical tools
- General d-i-y tools

Connections within round rear-entry conduit box

Start by choosing the type of fitting you want to install. The cheapest is a simple bulkhead light, but you can choose from a wide range of more ornate fittings which look attractive by day too, or go for high-level floodlighting if you want to illuminate a large drive or parking area. Floodlights usually need a lot of power (500W is typical) so make sure the fitting you buy will not overload the circuit you are adding it to. Make sure that any fittings you choose are suitable for outdoor use.

Next, work out where to position the light on the house wall. The best way of doing this is to wire the fitting temporarily to an extension cable and get a helper to hold the light against the wall after dark in various positions (using steps or a ladder if necessary) while you check the effect. If you are lighting a stepped path, make sure that each tread is well lit for safety. Although a little light goes a long way in the dark, you may find that you need more than one fitting to light your approaches properly.

Decide how you are going to provide power to the new lights. The easiest method, involving the least disruption, is to take a spur from a power

circuit socket outlet (see Option 1). Alternatively, you can extend a lighting circuit (Option 2), but this may mean lifting floorboards to track down where the circuit cables run. The ground-floor lighting circuit will be the most convenient to use unless you are installing high-level floodlighting, which will be easier to connect to the first-floor circuit. With lighting circuit extensions, check that the new light(s) will not overload the circuit. Also check that the circuit is designed to disconnect within 0.4 seconds in the event of a fault – a professional electrician will be able to advise you – and fit an RCD if not.

Check whether the fitting you are installing has a hollow baseplate that will accommodate the wiring connections. If it has you can mount the fitting straight on the wall with the cable emerging through the wall behind it. If it does not, you will have to fit a recessed round rear-entry conduit box in the wall for the connections, and then mount the fitting over it.

2 Drill a hole in the house wall at the point where the light will be fitted, using a large masonry drill long enough to penetrate its thickness. Drill through both leaves of a cavity wall in one operation. Angle the hole slightly upwards from the outside to discourage rainwater penetration. Then line it with a length of PVC conduit to prevent the cable chafing on the masonry, and feed it through from inside the house. If you are using a conduit box, solvent-weld the joint between conduit and box.

3 Strip the cable ends, prepare the cores and connect them to the fitting. Some fittings have a terminal block; others have a short length of flex attached, and in this case you must connect the flex and cable cores together using strip connectors. Cover the cable's bare earth core with green/yellow PVC sleeving before connecting it up. If you are using strip connectors, wrap the completed connections carefully in PVC insulating tape to keep out the damp.

Alternatively, push the wired-up strip connector into the fitting's baseplate or into the conduit box, and pack silicone mastic round it.

4 Draw the cable back into the house and screw the fitting to the wall or box. Then seal round the edges of its baseplate with a bead of silicone mastic to prevent water from getting in.

5 Inside the house, what you do depends on how you intend to run the wiring:

Option 1:

If you plan to take a spur from a socket outlet circuit to a switched fused connection unit (FCU) and intend to use this as the on-off switch for the light, decide on a convenient position for it and run the 1mm² cable from the light to this position – ideally in vertical and horizontal chases cut in the plasterwork, but surface-mounted if you prefer.

Connect the cable to the LOAD terminals of the FCU. Then run 2.5mm² cable from the FEED terminals of the FCU to a suitable nearby socket outlet (see pages 92-3), ready for connection to its terminals. Fit a 5-amp fuse in the FCU.

If you want separate switch control, run the cable from the light to a four-terminal junction box first. Then run another 1mm² cable from the junction box back to the FCU, and connect it to the LOAD terminals as before. Connect the switch cable into the junction box and run it to the chosen switch position. You can, of course, have additional two-way switches to control the lights if you wish, perhaps siting a weatherproof one outdoors and one indoors for convenience.

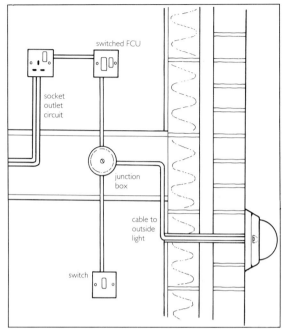

Below: outside lights fed from a socket outlet circuit (top) and a lighting circuit (bottom)

Option 2:

If you prefer to supply the outside light from an existing lighting circuit, locate the circuit cable and run the cable from the light to a four-terminal junction box which will be cut into the circuit at a convenient point. Then run a length of 1mm² cable from here to the switch position, and connect it up. Flag the neutral cores at each end of the switch cable with red PVC tape to indicate that they are live.

6 It is now time to make the mains connections. Turn off the power to the circuit concerned at the consumer unit and check that the circuit is dead. Then remove the socket faceplate and connect in the spur from the FCU to complete a socket outlet circuit extension, or cut the lighting

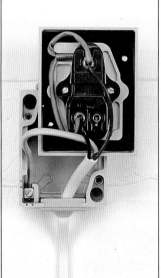

circuit cable next to the four-terminal junction box, and connect them to the box terminals for a lighting circuit spur.

Double-check all the connections and the continuity of

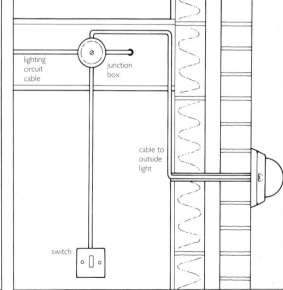

the conductors on your new sub-circuit, then restore the power and test the light.

Above left: weatherproof switch – remember to flag the black core

RUNNING CABLE OVERHEAD

If you want to extend your electricity supply beyond the house walls, you can run the circuit cable in one of two ways: overhead or underground. The former is easier and quicker to install, but the latter is far less obtrusive and is also safer in the long run.

Overhead wiring is a popular and economical way to provide a power supply to an outbuilding, especially if it is relatively close to the house. It is not really practical for wiring to garden lights or outdoor socket outlets, which are normally given an underground supply (see pages 126-7).

Plan out the cable run, checking spans and ground clearances so you can decide where the cable should leave the house, how it will reach the outbuilding and whether a catenary wire is needed. Attach a support post to the outbuilding if necessary to achieve adequate clearance.

THINGS YOU NEED

- 2.5mm² two-core-and-earth cable for 20-amp circuits
- 4mm² cable for 30-amp ones
- 16mm round PVC conduit
- 20mm steel conduit, saddles and PVC bushes plus earth cable and connector (optional)

For runs over 3m
- Galvanized catenary wire
- Eyebolt fixings
- Tensioner
- Cable buckles and clips
- Earth cable connector
- 4mm² single-core earth cable
- Support posts if required

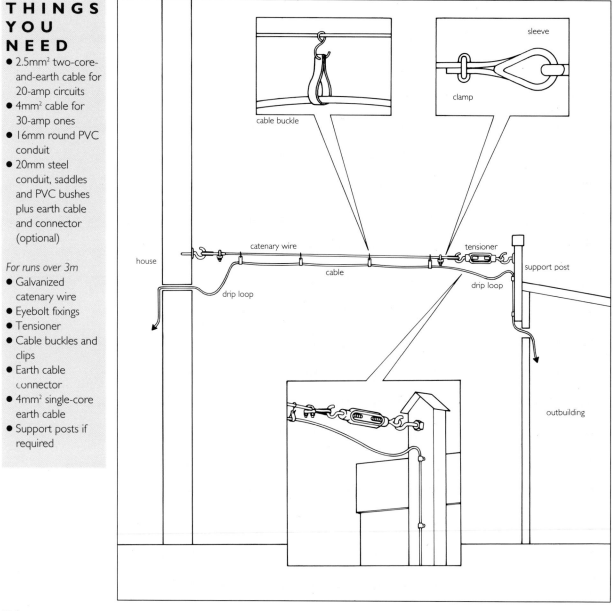

sleeve

clamp

cable buckle

catenary wire

tensioner

house

cable

support post

drip loop

drip loop

outbuilding

2 If a catenary wire is needed, fit an expansion anchor with an open eyebolt at the house end of the cable run and attach a tensioner to it. Fit another eyebolt to the outbuilding or the support post, and attach one end of the catenary wire securely to it. Draw the other end of the wire through the tensioner and pull it as tight as possible by hand and clamp it. Then turn the tensioner with a screwdriver to pull the wire taut.

3 Drill a hole through the house wall (just below the eyebolt position if a catenary wire is being used), angling it slightly upwards to discourage water penetration, and insert a length of 16mm round PVC conduit to prevent the cable from chafing. Make an entry hole at the outbuilding, again sleeving it if it passes through masonry.

4 Unroll the circuit cable and feed one end in through the hole in the house wall. Take it back to the consumer unit position by the most convenient route, leaving ample slack there for the final circuit connection to be made.

5 For spans of less than 3m, simply draw the cable across from the house to the outbuilding and secure it to the support post there with cable clips.

Check that the ground clearance is adequate, then feed the cable into the outbuilding ready for connection to its electrical equipment (see page 133).

6 If steel conduit is being used to provide additional protection for the cable span, secure the length (a maximum of 3m; 10ft) to the buildings or support posts with conduit saddles. Fit a PVC bush to each end of the conduit, feed the cable through it and take it into the building.

7 If a catenary wire is being used, form a drip loop where the cable emerges from the house wall. This is a 'droop' in the cable which allows rainwater to fall to the ground and keeps it away from the house wall. Then attach the cable to the wire at about 230mm (9in) intervals with cable buckles. When you reach the outbuilding, form another drip loop before clipping the cable down the support post and into the building.

8 Attach the earth cable connector to the free end of the catenary wire next to the tensioner or to the steel conduit. Then run a length of 4mm² single-core earth cable from the connector through the hole in the house wall and back along the circuit cable route to the house's main earthing point. Connect it to the earth terminal.

9 Complete all the wiring within the outbuilding (see pages 133) before making the final circuit connections inside the house.

Wiring regulations

Ordinary PVC-sheathed cable can be used for the entire run, and if the span between the house and the outbuilding is less than 3m (10ft) the cable needs no additional support so long as care is taken to prevent the cable from chafing where it passes through walls. It can, however, be given extra protection by being run in a single unjointed length of 20mm diameter steel conduit. The maximum span if conduit is used is again 3m, and the conduit ends must be fitted with PVC bushes to prevent the cable from chafing on the metal. The conduit itself must also be earthed.

If the span is longer than 3m, the cable must be supported by a catenary wire to which it is bound by suitable buckles. The catenary wire is a multi-stranded galvanized steel wire, available from electrical suppliers, which is secured to stout eye bolts at each end of the span. Fitting a tensioning device at one end makes it easier to pull the wire taut and prevent sagging. The wire must be earthed to the house's main earthing point with a length of 4mm² single-core earth cable.

Both unsupported and supported overhead cables must have adequate ground clearance to prevent them from being damaged accidentally. Where there is only pedestrian traffic, the minimum clearance allowed is 3.5m (11ft 6in), or 3m (10ft) if steel conduit is used; this must be increased to 5.2m (17ft) over driveways and other areas with vehicle access, and conduit is not allowed in these locations.

On long runs, intermediate supports will have to be provided for the cable and the catenary wire, in the form of posts tall enough to give the necessary ground clearance. This is when overhead cable runs become obtrusive (and also more expensive), since you are not allowed to extend existing fence posts to carry the cable; it must have independent supports.

RUNNING CABLE UNDERGROUND

By far the neatest and safest way of running cable out of doors is to bury it underground, out of harm's way. This is obviously more time-consuming (and more expensive) than using overhead wiring, but is well worth the effort in the long run.

Underground cable runs can be installed in two different ways. If armoured or metal-sheathed cable is used, the run can be buried directly in the ground. If ordinary PVC-sheathed cable is used, it must be protected along the entire length of the run by being enclosed in impact-resistant PVC conduit, assembled with waterproof solvent-welded joints.

Armoured and metal-sheathed cables are expensive. Both are used only for the underground section of the circuit, and have to be connected to the indoor circuit wiring using special junction boxes and screwed glands. This is a fiddly job with armoured cable, although well within the capabilities of the average do-it-yourself electrician. However, the cut ends of metal-sheathed cable have to be fitted with special seals to prevent the mineral insulation inside from absorbing moisture, and since this requires specialist tools, using this type of cable is generally not recommended for the amateur.

Because of both the expense and the fitting complications involved in using these two cables, the cable-in-conduit method is generally preferred for domestic underground cable runs. The job is simpler and the extra cost of the conduit and its fittings is offset by the savings on cable, junction boxes and special glands and seals; the circuit cable can run continuously from source to point of use without a break.

Unlike an overhead cable run, an underground one has to be protected along its entire outdoor section. This means that the conduit passes through the house wall, turns down it via an elbow if it exits above ground level, runs down to the bottom of the trench and then turns through another elbow to run along the trench bottom to its destination, where a final elbow brings it to the surface again. The elbows and any couplers used to join the conduit lengths must be solvent-welded as the run is assembled to prevent moisture from entering the conduit. The section of the run that is buried underground must be fully assembled (with a draw wire inside) before the cable is drawn through it. This section should be a straight run, since drawing cable round the small-radius elbow fittings is difficult. Then the above-ground conduit and fittings are fed on to the cable before being solvent-welded in place.

The depth at which the run is buried must be sufficient to prevent it being disturbed. A depth of 450mm (18in) is adequate beneath paths and other paved areas, but should be increased to 750mm (2ft 6in) beneath lawns and flower beds which could be dug up in the future.

To warn future diggers of the presence of the cable, it must be identified with special tape laid over it, and can be given extra protection with roof tiles, paving slabs or proprietary covers.

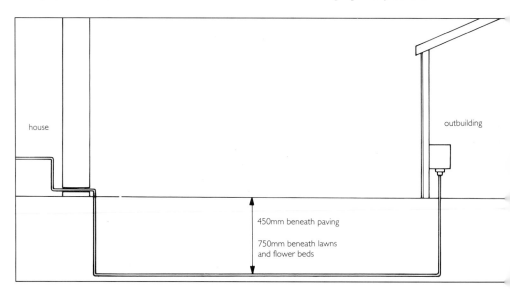

house

outbuilding

450mm beneath paving

750mm beneath lawns and flower beds

1 Plan the line of the cable run carefully to minimise the amount of excavation work necessary, especially if the run has to cross paved areas. It is possible to tunnel under narrow paths without having to dig them up.

2 Drill the exit hole for the cable at a convenient point on the house wall, and insert a length of conduit through it. Then start digging the trench immediately below the hole. If the trench runs to an outbuilding, drill an entry hole in its wall above the end of the trench.

3 Assemble and solvent-weld the underground section of the conduit run, with a draw wire inside it, and lay it in the trench. Then draw the circuit cables through it.

4 Feed lengths of conduit and fittings on to the house end of the cable run to complete the conduit run there, and feed this end of the circuit cable in through the hole in the house wall. Take it back to the consumer unit position, leaving enough slack for the final circuit connection to be made later on.

5 Make up the outbuilding end of the conduit run in the same way, solvent-welding couplers and fittings to the conduit as you proceed. Use clips to secure the vertical section at the house wall, and to fix the other end of the run to the wall of the outbuilding, a garden wall or a support post as appropriate.

Using Armoured Cable

Armoured cable can be used unprotected in an underground trench. At each end of the run it is brought into the house or outbuilding and is connected to ordinary PVC-sheathed cable at a conversion box. Here a special metal cable gland and locknut (made to BS6121) is used to connect the cable to the box. An earth tag beneath the locknut earths the cable armouring. Then its cable cores are connected to the cores of the PVC-sheathed indoor cable run via strip connectors which are housed within the conversion box; this has a screw-on cover that is fitted once the cable connections have been made.

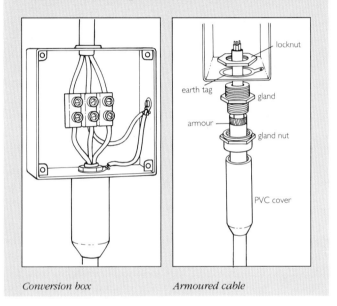

Conversion box   Armoured cable

THINGS YOU NEED
- 2.5mm² two-core-and-earth cable for 20-amp circuits
- 4mm² cable for 30-amp circuits
- Impact-resistant PVC conduit
- Solvent-weld cement
- Cover tape

6 If the run is serving several garden lights or socket outlets, fit an elbow at each light/outlet position so the circuit cable can run up to the fitting, then back down again through a parallel length of conduit and an elbow to the next one via another underground section.

7 Lay cover tape over the conduit and backfill the trench. If you want to give the conduit additional protection, shovel in about 150mm (6in) of backfill first and tread it down carefully. Then lay roof tiles, slates or pieces of paving slab on top before backfilling the rest of the trench.

8 Complete all the wiring work at the 'downstream' end of the circuit before making the final circuit connections within the house.

FITTING REMOTE OUTSIDE LIGHTS

Lighting up the garden, as opposed to fitting 'task' lighting to help you find your way to the front door, not only increases the amount of time you can spend outdoors after dusk; it can also create some spectacular effects for you to enjoy into the bargain.

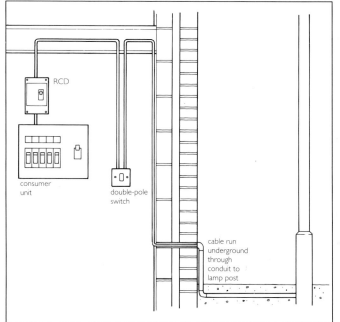

Install remote outside lighting on its own permanent circuit or take the power from outdoor sockets as required

The first thing to remember when you are planning garden lighting is that a little light goes a long way in the dark – a 25-watt bulb used outside will seem far brighter than it would indoors. Next, you need to be aware of the effect of stepping into or out of a brightly lit area; you can be dazzled by badly positioned lights, and as you step out into the darkness you can find yourself virtually blind for a few seconds until your eyes adjust to the different light level, especially if spotlights are being used. Lastly, use coloured lights with care, especially if you are illuminating trees and shrubs; they tend to create some rather weird colour effects.

Before beginning to think about what types of lights to use in your garden, there are two other decisions to make. The first is whether to use mains-powered or extra-low-voltage lights. The former allow you to achieve a greater variety of effects, but they will be more expensive and time-consuming to install because a separate lighting circuit must be run from the house and buried underground. The latter are safer to use because they operate at only 12 volts, and they are much simpler to install since you can just run

the cable on the ground from the transformer to the various lighting positions; however, they give you rather less scope for any dramatic lighting arrangements. See page 131 for more details.

Think also about flexibility: you have to decide whether the lights will be needed at fixed points – round a barbecue area, for example – or whether you will want to move them around to highlight different features of the garden as the seasons change. If you decide you need flexibility, you may have to plan a circuit that provides weatherproof socket outlets at various posi-

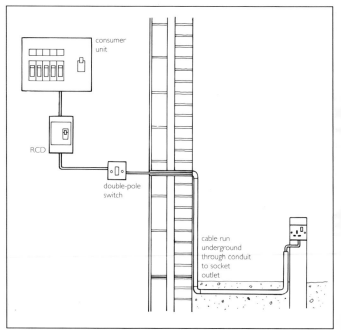

ons, allowing lights to be disconnected and reconnected at
will; alternatively, stick to
extra-low-voltage types which
can be lifted and repositioned
as required.

There is a huge range of
outdoor light fittings available,
from free-standing lamp posts
and low-level bollards (ideal
for lighting a garden path, for
example) to globes, lanterns
and flood- and spotlight fit-
ings, often supplied with a
ground spike fixing. If you are
looking for permanent fixtures
and there is a wall or outbuilding close at hand you can also
use any outdoor wall-mounted
type. Note that mains-voltage
lights and cables must not be
fixed to fences. When shopping, double-check that any
fittings you buy are marked as
suitable for exterior use.

1 For a mains-powered
installation, plan out the
circuit in detail. You have to
decide where individual lights
or socket outlets will be positioned in the garden, work out
how to route the underground
cable runs and provide a separate circuit from the main
house consumer unit. If you
do not have a spare fuseway
available, you will have to add
a new small consumer unit.

2 Dig the trenches to each
light/socket outlet position
and install the cables (see
pages 126-7). If socket outlets
are being used, set up short
stout timber posts in concrete
and clip the conduit to them.

3 Mount the fixed lights or
weatherproof socket outlets
as required, then strip and
connect in the cable ends.
Double-check that fittings or
socket outlets are properly
earthed, and that sealing grommets are fitted carefully to

exclude moisture. With socket
outlets, plug in the lights at
their chosen positions; check
that the waterproof covers are
correctly fitted to any unused
outlets.

4 Make the final connections
at the consumer unit.
Install a high-sensitivity
(30mA) RCD at the start of the
circuit to provide extra electrical safety; this is mandatory to
satisfy the Wiring Regulations
if the circuit is supplying socket outlets rather than fixed
lights. If you are adding a new
consumer unit, it can incorporate the RCD alongside the
new circuit fuse or MCB; if you
are using an existing spare
fuseway, position the RCD in
its own enclosure close by and
wire the new circuit cable
through it. Use a 5/6-amp fuseway for a circuit to fixed lights,
and a 20-amp one for circuits
to socket outlets.

5 You can use the RCD as
the main on-off switch for
the circuit, but it is more convenient to have a conventional
wall-mounted switch some-
where – by the back door, for
example – to control the circuit. For both types of circuit,
use a double-pole switch to
provide complete isolation
when the lights/socket outlets
are not in use.

*Left: Protect outdoor
lighting with an in-line
RCD. This can be used
as the on-off switch for
the lights*

AUTOMATIC LIGHTING CONTROL

While you can turn your outside lights on and off using conventional switches, you can improve their efficiency as security devices by putting them under the control of an automatic switch – either a photoelectric type which turns the lights on and dusk and off at dawn, or else a passive infra-red (PIR) detector which operates them when it detects someone in its field of vision.

THINGS YOU NEED

- Outside lights
- Photoelectric switch or PIR detector
- 1mm² two-core-and-earth cable
- Green/yellow PVC earth sleeving
- Conduit and fittings (if required)
- Double-pole isolating switch (if required)
- Electrical tools
- General d-i-y tools

Above right: Outdoor switches and detectors

Right: PIR detector. Flag the black cores with red PVC tape before mounting the switch

PHOTOELECTRIC SWITCH

A photoelectric switch incorporates a time-delay mechanism to prevent it from being activated by short-term light conditions such as the headlights of a passing car. It should be installed on a north-facing wall so it is not in direct sunlight, and should not be able to 'see' the lighting it is controlling. A typical switch can control up to 1,200 watts of tungsten lighting or about 500 watts of fluorescent lighting, and can be overridden manually if an ordinary switch is wired into its external switch and neutral terminals.

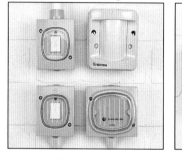

PASSIVE INFRA-RED (PIR) DETECTOR

A passive infra-red detector switches the lights it controls on whenever it detects movement within the area monitored by the sensor, and turns them off again after a pre-set time if the heat source leaves the area or its movement ceases. For maximum efficiency the sensor should ideally be mounted so that people or cars will cross its detection zone rather than approach it head on, and should be fitted about 2.5m (8ft) above ground level. The sensor may be a separate unit (which can typically control up to 2,500 watts of tungsten lighting), or may be combined with a light fitting as an integral unit. Sensors can also be wired in series to increase the detection area if required.

1 To fit a separate sensor, mount its backplate at the chosen location, then attach the sensor to the backplate. Install a conduit run to it if required.

2 Install the lights to be controlled, and run cable from them back to the sensor position. Then run the lighting circuit cable from its supply point to the sensor, via a double-pole isolating switch if required by the installation instructions.

3 Follow the sensor manufacturer's installation

1 Fit the switch mounting box at the chosen control position, and install a conduit run to it if required.

2 Install the lights to be controlled, and run cable from them back to the switch position. Then run the lighting circuit cable from its supply point to the switch, via a double-pole isolating switch required by the installation instructions.

3 Connect the circuit cable to the switch live, neutral and earth terminals, and the light cable to the external load neutral and earth terminals.

4 Check the operation of the unit as described in the switch manufacturer's installation instructions, then secure the switch to its mounting box.

Left: Photoelectric switch. Flag the black cores with red PVC tape before mounting the switch

instructions for wiring the unit continuous monitoring is the usual mode, but other variations are possible.

4 Check the operation of the sensor as described in the instructions, set the required pre-set time and fit the sensor to its mounting box.

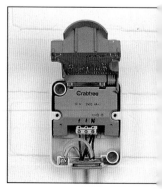

The major advantage of using extra-low-voltage lighting in the garden is that there is no need to protect the circuit cable by running it in conduit or burying it underground. Instead it can simply be run on the surface – across flower beds, for example – to the light positions. This allows you to move the lights around if you wish; they are generally mounted on spikes which you simply push into the ground.

The main disadvantage of this type of lighting is that the maximum cable length is restricted by the relatively high current it carries, so unless you use several separate circuits (each with its own transformer) you cannot illuminate a very large area. A typical garden lighting set includes four lights plus about 8m (26ft) of cable.

The lights themselves are available in several styles, including lanterns, bollards and reflectors, and there are special submersible or floating types designed for use in garden ponds. The transformer, which can supply only a fixed number of lights, must be sited indoors, in an outbuilding or in a suitable weatherproof enclosure.

1 Start by deciding where you want the lights, then lay out the cable to run from the furthest light back via the others towards the transformer position.

2 Assemble the light fittings as described in the manufacturer's instruction leaflet. Then unscrew the terminal cover plate to expose the terminal pins inside.

3 Double-check that you are positioning the light precisely where you want it on the cable run, then press the two-core cable on to the terminal pins so that one pin pierces the insulation on each core and makes contact with the conductors inside. Replace the terminal cover and position the lights by pushing their spikes into the ground.

4 Connect the circuit cable to the output terminals of the transformer. Then fit a plug with a 3-amp fuse to its

FITTING EXTRA-LOW-VOLTAGE OUTSIDE LIGHTING

Mains-voltage lighting is perfectly safe so long as it is carefully installed, especially if the circuit also has RCD protection, but can be expensive. An alternative is to use extra-low-voltage lighting, run from a transformer, although this does not have the versatility of mains-voltage lighting.

flex and plug it into a switched socket outlet. This will also act as the on-off switch for the lights. If the transformer is sited outside the house, either in an outbuilding or in a weatherproof enclosure, the socket supplying it must be protected by a high-sensitivity (30mA) RCD.

THINGS YOU NEED

- Extra-low-voltage lighting set(s)
- Screwdriver
- Plug with 3-amp fuse

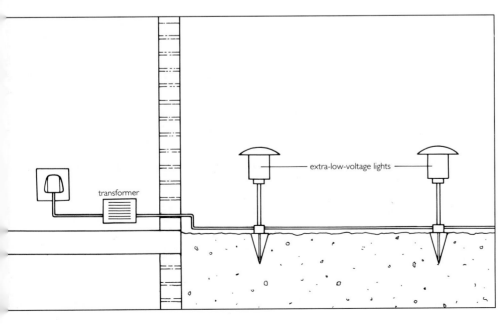

transformer

extra-low-voltage lights

FITTING OUTDOOR SOCKET OUTLETS

If you use a lot of garden power tools, you will soon get tired of trailing flexes out of the kitchen window to power them. The solution is to install one or more socket outlets purely for garden use.

THINGS YOU NEED

- Weatherproof socket outlet with RCD or socket outlet and separate RCD
- 2.5mm² two-core-and-earth cable
- Green/yellow PVC earth sleeving
- 16mm round PVC conduit

Below: Circuit to outdoor socket outlet

If you want the convenience of having several socket outlets dotted around the garden, follow the instructions given on pages 128–9 for installing outlets for outside lighting, using 2.5mm² circuit cable and a 20-amp circuit fuse or MCB. Add a high-sensitivity RCD to the circuit at or next to the consumer unit to satisfy the Wiring Regulations requirements for circuits supplying outdoor socket outlets.

If all you want is somewhere to plug in the lawnmower, you can simply install a special weatherproof socket outlet on the house wall, wired up as a spur from the downstairs power circuit. To satisfy the requirements of the Wiring Regulations this socket must, like an outdoor socket outlet circuit, be protected by a high-sensitivity RCD, and the simplest way of providing this is to fit an outlet with a built-in RCD. Alternatively, use an ordinary weatherproof socket outlet and wire it up via a separate RCD on the spur cable.

I Start by locating a suitable socket outlet as the source of the spur – one on the ring circuit, but not already supplying a spur (see pages 92–3). Drill a hole through the house wall at the point where you want the new socket outlet, line it with a length of conduit and feed in the spur cable.

2 Pass the cable into the socket's mounting box via a weatherproof grommet and secure the box to the wall. Then prepare the cable cores, connect them to the socket terminals and mount the socket on its box. Check that all weatherproof seals are correctly placed.

3 If you are using an outlet with an integral RCD, turn off the power to the socket outlet you plan to use as the source of the spur at the consumer unit and check that the circuit is dead. Then undo its faceplate, feed the spur cable into the mounting box and connect the cores to the terminals. Replace the faceplate, restore the power and test the installation.

4 If you are using a separate RCD, run the spur cable back to the RCD (which can be wall-mounted in its own enclosure next to the existing socket), connecting the live and neutral cables to its terminals and linking the earth cores to the earth terminals. Then run another length of cable from the RCD to the existing socket outlet and make the connections as in Step 3.

Below: Weatherproof socket outlet

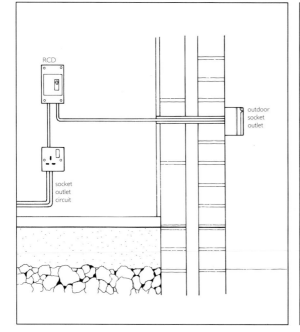

From the house, the cable is run overhead (see pages 124–5) or underground (pages 126–7) to the outbuilding, where it is connected via an RCD to another small consumer unit or a double-pole switch. This allows the building's power supply to be isolated from within the building as well as in the house.

Within the outbuilding, it is best to provide separate circuits for the light(s) and the socket outlets. The first is wired from a 5/6-amp fuseway in the consumer unit, the second from a 20-amp one. Alternatively you can have a single 20-amp circuit, with the lighting fed via a spur to a fused connection unit (FCU) which is fitted with a 5-amp fuse. It is a good idea to use durable metal-clad wiring accessories in the building, rather than standard white plastic ones, and you can also if you wish run the cables in round PVC conduit.

Wiring regulations

Since the Wiring Regulations insist on RCD protection for all sockets in the building, you must install a high-sensitivity (30mA) RCD to protect the entire outbuilding sub-circuit. This can be within the outbuilding consumer unit. You have to earth the outbuilding installation with 6mm² earth cable to an earth rod driven into the ground next to the building.

Top: outbuilding wiring using two separate circuits. An alternative method is shown on page 52

Far right: The wiring within the new consumer unit and metal-clad wiring accessories

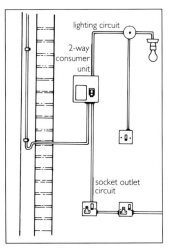

1 Bring the circuit cable into the outbuilding and run it to the consumer unit or DP switch position via the RCD. Do not connect the earth cable to the outbuilding circuit. Earth the installation with 6mm² single-core earth cable to an earth rod driven into the ground next to the building.

2 If installing just one circuit, run 2.5mm² cable to the FCU and connect it to the FEED terminals and the earth terminal. Run 1mm² cable from the FCU's LOAD terminals to the light fitting, and connect it to its terminal block. Run more 2.5mm² cable on from the FCU's FEED terminals to the socket outlet. If more than one outlet is fitted, wire them up as a radial circuit.

3 If installing separate circuits to the light(s) and the socket outlet(s), run each one in the appropriately sized cable from its fuseway or MCB in the consumer unit. Take the lighting circuit to a four-terminal junction box, then add cables to the switch and light. Alternatively, you can dispense with the junction box and light fitting; instead use a loop-in battenholder. Wire the socket outlets as a radial circuit.

WIRING IN OUTBUILDINGS

If you want to provide lighting and power for a garden shed or workshop, you must install a complete sub-circuit; you cannot just extend your house lighting and power circuits. This means using a 30/32-amp fuseway in your consumer unit if you have a spare one, or installing a separate unit next to it otherwise.

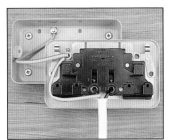

THINGS YOU NEED

- Small consumer unit or double-pole switch
- High-sensitivity RCD
- Metal-clad wiring accessories
- Metal surface mounting boxes
- Four-terminal junction box
- Light fitting
- 2.5mm² two-core-and-earth cable
- 1mm² two-core-and-earth cable
- 6mm² single-core earth cable
- Green/yellow PVC earth sleeving
- Conduit to protect cables (optional)
- Earth rod and clamp
- Electrical tools
- General d-i-y tools

REPAIRS & MAINTENANCE

If something electrical stops working, many people simply throw in the towel and either call an electrician or take the appliance concerned back to the repair shop. However, in many instances the fault is simple both to detect and to put right, often at minimal expense.

The golden rule when investigating any electrical fault is to isolate whatever you are examining from the mains supply. This means unplugging appliances from socket outlets, turning off double-pole switches where these control the supply to fixed electrical equipment, and removing circuit fuses or switching off miniature circuit breakers (MCBs) before touching anything else. Do not restore the power supply again until you have either traced and corrected the fault or you have isolated the offending piece of equipment from the system.

The Problem Checklist on pages 8–10 outlines how to go about tracing faults of various types, and tells you briefly what to do to put things right. The information in this section tells you more about what causes electrical faults, and explains in more detail some of the techniques you will need to tackle them. However, you should not attempt any repairs that are beyond your capability. In particular, many modern electrical appliances are extremely complex, and you can often do more harm than good (and multiply the eventual repair bill) by tampering with them; you may also invalidate any guarantee in force.

Remember too that the incoming supply cable to your property, the service fuse (cut-out) and the electricity meter all belong to your electricity supply company, and it is a legal offence to tamper with them. You must also not interfere with the meter tails leading to your consumer unit. Leave any work involving them to a qualified electrician.

WHAT CAUSES FAULTS

The commonest causes of electrical faults are loose connections and broken conductors. They generally occur on portable appliances or within their plugs, but can also affect pendant light fittings. If a conductor core breaks or becomes completely detached from its terminal, the appliance or light will simply stop working. More often than not, however, the conductor remains in partial contact with its terminal and although the appliance still works, the higher-than-normal impedance generates heat which can do further damage and which may start a fire.

Another problem a loose connection can cause is a short circuit. If the loose conductor then touches another terminal, the current bypasses the appliance or light, which again stops working. The resulting surge in current also blows a fuse, either in the appliance plug or at the consumer unit. The same thing occurs if the insulation in a cable or flex is damaged or breaks down, allowing live and neutral cores to touch.

If a loose connection or faulty insulation allows contact to occur between a live or neutral conductor and either the earth conductor or any earthed metalwork such as the casing of an appliance, there is again a current surge, this time as electricity flows to earth, and the surge again blows a fuse (or trips an RCD). If the contact is with metalwork that is not properly earthed and you touch the metal, you will get a shock as the current takes the only available path to earth: through your body.

The other common fault is overloading a circuit; that is, trying to draw more current than it is designed to deliver safely. So long as the circuit is properly protected against overloading with an appropriate fuse or MCB, the protective device will cut off the current flow before any serious damage is done. But if the fuse has been replaced with one of too high a rating for the circuit, or if some other metallic object has been fitted in its place, excess heat will be generated which will cause damage to insulation and will possibly start a fire.

rcd tripping

If an RCD trips off, your first action should be to attempt to reset it.

- *If you can, the fault was an intermittent one which may not recur (if it does, get professional help). If you cannot reset it, disconnect all appliances, then reset it and reconnect the appliances in turn to identify the culprit.*
- *If you cannot reset it after disconnecting all appliances, remove all the circuit fuses or switch off the MCBs. If you can now reset it, there is a live-to-earth fault somewhere on the system.*
- *If you cannot reset it, disconnect all the circuit neutral cores at the consumer unit. If you can now reset it there is a neutral-to-earth fault.*

mcb's

MCBs have two advantages over wire fuses; their convenience in use, and the fact that they cannot be abused by being replaced with other metallic objects, as fuses can. However, there is one electrical fault which causes the latest type of MCBs to trip off but which does not apparently blow rewirable or cartridge fuses (or older 'half-cycle' MCBs). When a filament lamp fails, pieces of the element can cause a short circuit within the lamp envelope. Good-quality lamps contain a fuse in the base that is designed to blow if this occurs; in lamps without one, the short circuit can trip the lighting circuit MCB. This is an obvious nuisance since it will black out part or all of the house. If you have MCBs and have experienced this problem, the solution is to install less sensitive MCBs on the lighting circuits.

TRACING FAULTS

Since electrical faults are for the most part localised malfunctions, it follows that the secret of success in tracing them once a fuse has blown or an MCB has tripped off lies in methodical detective work.

From top: Use a continuity tester to check plug fuses, the continuity of individual flex cores and appliance components such as switches

Often the cause of the fault is clearly visible: a loose connection at a terminal, for example, or a hole drilled through a buried cable. Sometimes it is not, and you will need to use some simple test equipment to trace the site of the fault by a process of elimination.

I If an appliance stops working with no visible signs of a fault – a bang as the plug fuse blows, or the smell of overheating – start by unplugging it and checking the plug itself. If the connections are sound and properly made, check the plug fuse next.

2 Then check the continuity of each core between the plug and the appliance terminal block. Test the wiring between there and the appliance's on-off switch, and finally check the switch itself for continuity.

3 Even if you find no fault and can go no further, you will at least have eliminated the most obvious culprits and narrowed the site of the fault to within the appliance itself. Use the same principles to track down faults on your house circuits, after turning the power off at the mains.

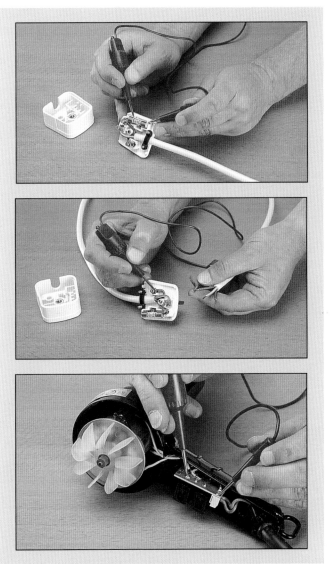

REPLACING FUSES

On a well-designed and well-maintained wiring system, fuses will blow seldom if at all. However, even if you have MCBs in your consumer unit you should still keep a supply of spare plug fuses in the house; if you have circuit fuses, keeping the appropriate spares is essential.

THINGS YOU NEED

- Replacement plug fuses
- Fuse wire or cartridge fuse to match circuit rating
- Continuity tester
- Screwdriver

Below: Step 3
Below right: Step 4

If a fuse in a plug or fused connection unit (FCU) blows, carry out the usual checks to establish why it blew and correct the fault before fitting a replacement fuse and operating the appliance again. Always ensure that fuses are marked 'made to BS1362' or carry the ASTA mark, and fit a 3-amp fuse to appliances rated at up to 720 watts, a 13-amp one otherwise. For colour televisions, check the manufacturer's instructions about fusing; some sets take less than 3 amps when running, but demand a higher starting current and so may need a fuse of a higher rating. You can if you wish fit 5-amp fuses to appliances rated at between 720

and 1200 watts.

If a circuit fuse blows or an MCB trips off, again attempt to trace the cause before replacing the fuse or resetting the MCB. If you have rewirable fuses, it is worth buying a spare wired-up fuseholder to match each of your circuit fuse ratings, so that you have a replacement ready to use whenever necessary; you can then rewire the blown fuse at your leisure. Similarly, if you have cartridge fuses, keep at least one of each rating, and replace any that you use so a spare is always available.

Remember that some other wiring accessories apart from FCUs also contain fuses. For example, many dimmer switches contain a cartridge fuse smaller than those used in plugs; these are made to BS646 and come in 1, 2, 3 and 5-amp ratings. Check this fuse if a light controlled by the dimmer stops working and the fault is not cured by replacing the lamp. Shaver socket outlets – the type used in rooms other than bathrooms and washrooms – also contain a BS646 fuse, which is usually rated at 1 amp.

I If you suspect that a plug fuse has blown, use a continuity tester to check it. If it has, trace the cause and correct the fault before fitting a replacement fuse of the correct rating for the appliance concerned.

2 When a circuit fuse blows, turn the power off at the mains and again trace the cause and correct the fault. Then remove the fuseholder from its fuseway.

3 To rewire a rewirable fuse, loosen the terminal screws and remove the remains of the old fuse wire. Fit new wire of the appropriate current rating between the terminals, leaving a little slack, and tighten the terminal screws. Snip off any excess wire, replace the fuseholder and restore the power.

4 To replace a cartridge fuse, open the fuseholder if necessary to gain access to the fuse. Fit a replacement – all are different sizes except for 15- and 20-amp fuses – and reassemble the fuseholder. Restore the power.

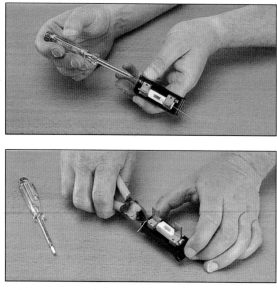

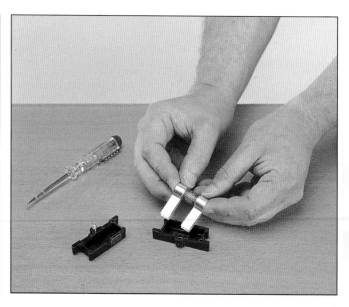

It is a wise safety precaution to check the condition of the flex on all your appliances regularly. The job will take only a few minutes to carry out, but will help you to spot potential problems before they become serious. Inspect the flex carefully, looking for any signs of damage to the outer sheathing, and if you find any make a temporary repair using PVC insulating tape. Do not regard such a repair as permanent. Also check the condition of the earth core using a continuity tester; a break in it will not stop the appliance from working, as a live or neutral break will, but will leave the appliance dangerously unearthed.

At the same time, check the appliance plug. Make sure the plug casing is intact, that the flex sheathing is securely held by the cord grip with no inner cores visible outside the plug, and that the terminal screws are tight. With portable appliances in particular, vibration can cause both terminal and cord-grip screws to work loose, and a regular – say half-yearly – check will prevent this leading to short circuits or other faults.

Always use the correct type of flex for the appliance concerned. Two-core flex is used only on appliances that are double-insulated and are marked with the double square symbol. The flex size must match the appliance wattage:

● 0.5mm²	up to 720 watts
● 0.75mm²	up to 1.4 kilowatts
● 1mm²	up to 2.4 kilowatts
● 1.25/1.5mm²	up to 3 kilowatts

When replacing the flex on an appliance, take the oppor-

tunity to fit a longer flex if this will make the appliance more convenient to use.

1 Unplug the appliance from the mains, open its plug and disconnect the flex cores from its terminals. If the plug does not have sleeved live and neutral pins, you should replace it with a new one that has.

2 Locate the screws that secure the appliance casing; you may need a range of cross-point, hexagonal or Torx-pattern screwdrivers to undo them. As you remove each one, tape it to the casing close to its hole so you know where to put it back; it is not uncommon to find that screws of different sizes and lengths have been used at various points, and they will not be interchangeable.

3 Note how the flex is routed into the appliance and locate the terminal block or switch to which it runs. Make a sketch showing which flex core goes to which terminal, then loosen the cord grip (if fitted) and the terminal screws, and withdraw the old flex.

4 Cut the new flex to length and prepare its cores, using the old flex as a pattern. If you are using braided flex, cut the braiding back neatly and wrap PVC insulating tape round the cut end to prevent it from fraying. Ensure that the earth core is cut long enough to remain slack and therefore connected to its terminal even if the other cores are pulled out of theirs by a sudden yank on the flex.

5 Feed the new flex into the appliance, via any grommet, cord grip or flex guide fit-

REPLACING APPLIANCE FLEX

Flex is surprisingly tough and can withstand not only normal everyday use but also a lot of misuse (and even downright abuse). However, if it does become faulty it can be highly dangerous and must be replaced immediately.

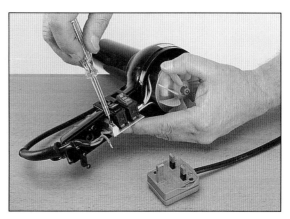

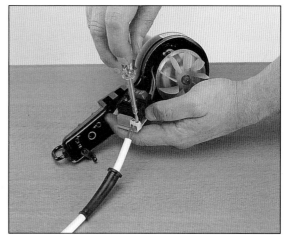

ted, and connect the cores to their terminals, checking your notes to ensure that each goes to the correct one. Tighten the terminal screws, secure the cord grip and close the appliance casing. Finally fit the plug to the other end of the flex. Discard the old flex.

THINGS YOU NEED
● Replacement flex
● Screwdriver
● PVC tape
● Electrical tools

EXTENDING FLEX SAFELY

The flex on many domestic appliances, especially portable ones, and also on power tools is often too short for convenience. Plugging the appliance into an extension reel is one solution, but it is often more convenient to lengthen the flex permanently.

If an appliance flex is too short to reach a socket outlet and span the distance to where you want to use it, you have three options. The first is to plug it into an extension reel. If you do this, always uncoil the reel fully first; otherwise it may overheat, melt the flex insulation and cause a short circuit. Take special care if using the reel to power appliances rated at over 1200 watts; many reels have flex rated at only 5 amps but may have a 13-amp fuse in their plugs, and so will overheat seriously if they are used to run appliances taking a higher current unless they are fitted with a safety cut-out.

The second option is to replace the appliance flex with a longer length (see page 137), but it may not be practicable to store the appliance with such a long flex attached, and you may find that you cannot open the appliance casing anyway to carry out the replacement.

The third option is to extend the existing flex. You must use a proper flex connector for this; twisting the cores together and wrapping the join in insulating tape is no way to do it –

it is a recipe for danger.

There are two types of flex connector. One-piece connectors are used to attach the extension flex permanently, while two-part connectors allow the extension to be attached when required and removed easily for storage. If several appliances – power tools and powered garden equipment in particular – are fitted with the same connectors, the lead can be used with all of them. Both types are available in different versions for connecting two-core and three-core flex. Always use flex of the same type and current rating as the original for the extension. For outdoor use, fit a splashproof toughened rubber connector and use white or orange flex.

Right: Wiring a one-piece connector
Far right: Two-part connector for powered garden equipment

1 Decide what extra length of flex you need and how you want to connect it to the appliance. Buy the appropriate type of connector and flex.

2 Remove the appliance plug and prepare the new flex cores. Lay the exposed cores over the connector and cut them to length so they will reach the terminals with the flex sheathing in the cord grip. Check whether you can reuse the core lengths on the existing flex, and shorten or lengthen them as necessary to fit the connector.

3 With a one-piece connector the brass strip connectors are usually loose, and it is easier to remove them to connect the flex cores. Connect like core to like core, then replace the strip connectors, secure each flex in its cord grip and fit the cover.

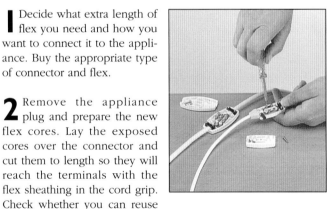

4 With a two-part connector, you must always attach the flex from the appliance to the male part of the connector – the one with the pins. If you attach the male part to the extension flex, the pins will be live when the power is on and could be touched if the connector was pulled apart.

With this type, you usually have to feed the flex through the cover and cord grip before connecting the cores to their

terminals. Then slide the cover up the flex and secure the terminal block inside it. With three-core flex, check that you wire each part of the connector correctly so that like cores are linked when the connector is assembled. It is vital that the earth cores are connected to the correct pins so that the earth connection is the first to be made when the two parts are joined and the last to disconnect.

Cables buried in walls are – or should be – run only vertically or horizontally, so you can work out how they reach individual wiring accessories and avoid drilling holes close to their likely routes. Cables run beneath floors should either be clipped to the joist sides or should pass through holes drilled in the joist centres so they are less likely to be damaged than if they are lying loose on the ceiling below or are in unprotected notches cut in the joist tops. Plastic conduit or channelling will offer little or no protection against a drill or saw.

If you drill into or cut through a cable, how you go about repairing it depends on where the cable is. If it runs beneath plaster, you will have to trace its run up and down the wall so you can replace the entire length. In a hollow stud wall or beneath a timber floor you should be able to rejoin the cable ends using a junction box. Switch off the power at the consumer unit and check that the circuit is dead before carrying out the repair, and mend the circuit fuse or switch the MCB back on when you have completed the work.

I Turn off the power at the consumer unit and check that the circuit is dead. If the cable is buried under plaster, chop this away as necessary to expose the cable run, disconnect it from any wiring accessory it serves within the room and cut through it within the floor void below or the ceiling void above as appropriate. Then join a new length of cable to the cut end using a three-terminal junction box in the void, and run the cable up or down the wall to the wiring accessory. Connect it to the accessory, secure it in the chase and plaster over it again.

2 If the cable is run in conduit or capping, you should be able to withdraw it from above or below after cutting through it and disconnecting it from the accessory it serves. Use the old cable to pull a draw cord through the conduit. then attach this to the new cable and draw this into place through the conduit. Connect it as in step 1.

3 If the damaged cable runs within a hollow stud partition wall, cut away a panel of plasterboard to expose the damaged cable. Then cut the cable at the point of damage, prepare the cut ends by stripping off the sheathing and core insulation, and rejoin them using a three-terminal junction box. Connect like cores to like cores, then screw the box to a stud in the wall and fit the cover before replacing the panel of plasterboard.

4 If the cable is beneath a suspended timber floor, lift a section of floorboard to gain access to it. As with stud walls, cut the cable and rejoin the cut ends using a three-terminal junction box. Screw this to the

REPAIRING DAMAGED CABLE
Because circuit cables are buried in walls and run under floorboards, there is a risk that they could be damaged during d-i-y work – usually by drilling or sawing. The result is generally a bang and a flash as the tool short-circuits the cable, plus a blown circuit fuse or a tripped MCB. If the circuit is also RCD-protected, the RCD will probably trip off too.

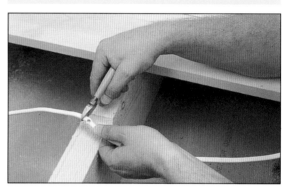

side of a nearby joist tools. If the cable was resting in a notch in the joists as shown, either screw a steel plate over the notch or re-route the cable through a hole drilled in the joist centre.

THINGS YOU NEED
- Three-terminal junction box
- Replacement cable
- Green/yellow PVC earth sleeving
- Electrical tools
- General d-i-y tools

REPLACING DAMAGED FITTINGS

Wall-mounted wiring accessories can be damaged by accidental impact, especially if they have plastic faceplates and are surface-mounted rather than fitted flush. If the faceplate or the box is badly cracked, live parts could be exposed and could give anyone using the accessory a shock. Swift replacement is essential.

THINGS YOU NEED

- Replacement accessory
- Replacement mounting box (if damaged)
- Screwdriver
- Terminal screwdriver
- Handyman's knife
- Cold chisel
- Hammer

If you crack or break a wiring accessory – when moving furniture, for example – your first priority is to make it safe. If possible, do not use it at all. Otherwise, and only as an emergency measure, cover the crack with several layers of PVC insulating tape; reinforce this with some stiff card if part of the faceplate is actually missing. If the damage is too severe to be repaired in this way and the accessory is in an essential location – the only light switch in the room, for example – you can get things working again by poaching a similar accessory from elsewhere in the house. Make sure the wiring is safe where the accessory has been removed. Buy a replacement at the earliest possible opportunity. It is a good idea to keep spare accessories for just this kind of emergency.

If the damaged accessory is in a particularly vulnerable location, it is worth considering fitting a replacement with a metallic faceplate rather than a plastic one, or turning a surface-mounted accessory into a flush-mounted one.

Ceiling roses and lampholders may need replacing for a different reason. Over the years the covers can become brittle and distorted due to heat rising from the lamp, and you may be unable to unscrew them to replace the pendant flex or to carry out wiring work at the rose itself. Rose covers can also become encrusted with paint, again making them impossible to remove. The only solution is to fit a new rose and lampholder.

1 Turn off the power to the damaged accessory at the consumer unit and check that the circuit is dead. If the accessory is flush-mounted or you have to replace a surface-mounted box, cut through any paint or wallcovering round the faceplate or box with a sharp knife. Then undo the faceplate fixing screws and ease it away from its box. Disconnect the cable cores after noting which go to which terminals, and discard the faceplate. Keep the fixing screws, though. Remove the mounting box too if this is damaged.

2 Fit a new box if necessary. Then reconnect the cores to the terminals on the new accessory faceplate, fold the cable(s) back into the box and fix the faceplate to it. If its fixing screws will not locate in the threaded lugs in the mounting box, try using the ones that held the old faceplate: the new ones may have a different thread pattern which does not match that in the lugs.

3 To remove a jammed ceiling rose cover, first turn off the power at the consumer unit and check that the circuit is dead. Then use a small cold chisel and hammer to crack the side of the cover so you can unscrew and remove it. Disconnect the circuit cable cores from the rose baseplate, unscrew it from the ceiling and discard it. Fit a replacement rose complete with a new pendant flex and lampholder, available as pendant sets ready-wired in various flex lengths. You could take this opportunity to fit a plug-in rose or luminaire support coupler rather than the traditional type.

Right: For an emergency repair, wrap several layers of PVC tape around the damage

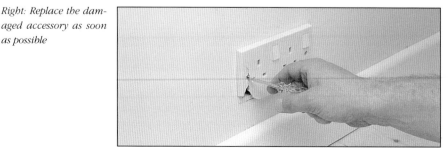

Right: Replace the damaged accessory as soon as possible

INSECURE FIXINGS

Wiring accessories rely on secure fixings to keep them in place – and safe. One type particularly at risk is the ceiling-mounted switch, where repeated and over-vigorous use of the cord pull coupled with insecure fixings can result in the switch pulling away from the ceiling.

1 Turn off the power to the switch at the consumer unit and check that the circuit is dead. Then unscrew the switch body from its baseplate. Remove the loose fixing screws and ease the baseplate away from the ceiling so you can inspect the fixing position and see what has failed. The baseplate should have been screwed either to the underside of a joist or to a batten fixed between the joists.

2 If the screws have simply pulled out of the timber, using longer screws may result in a better fixing. If the wood is badly splintered as a result

of the fixings failing, you may have to reposition the baseplate slightly to get a better fixing to the joist. Then reattach the switch to the baseplate and restore the power.

3 Similar fixing problems can affect wall-mounted socket outlets, especially those on hollow stud partition walls. Surface-mounted boxes can be pulled away from the wall, while flush ones may break away from the plasterboard. Turn off the power at the consumer unit and check that the circuit is dead before attempting a repair.

4 In the former case, remove the socket outlet faceplate so you can establish what type of fixing has been used. On solid walls, fitting new larger wallplugs and using longer fixing screws will cure the fault. On hollow walls, the box should have been screwed to a stud or to a horizontal nogging fixed between adjacent studs. Try using longer fixing screws.

OTHER REPAIRS

Apart from replacing damaged wiring accessories and repairing damaged circuit cables, there are one or two other repair jobs you may have to carry out to keep your system in good order.

5 If a flush hollow-wall mounting box has pulled out of the wall it will probably have damaged the edges of the cut-out in the plasterboard. Disconnect the faceplate from the supply cable(s), then remove the box. Check which way the cable runs so you can reposition the outlet in sound plasterboard a little further up, down or across the wall from the original position. Cut the new fixing hole, fit the box, feed in the cable and reconnect it to the faceplate. Patch the damaged cut-out.

REMOVING A BROKEN LAMP

If a lamp envelope is broken, or the envelope separates from the cap as you try to remove it for replacement, getting the cap out of the lampholder is a problem easily solved if you know what to do.

Turn the power off at the consumer unit and check that the circuit is dead. If the lamp is the bayonet-cap type, use a pair of pliers to grip the stem that carries the element support wires, then push the cap gently inwards and rotate it anti-clockwise to free the lugs from the lampholder. With Edison-screw lamps, grip the stem and turn the cap anti-clockwise to undo it.

TESTING YOUR WIRING

Any work you do on your wiring system must be carried out properly and safely. Working carefully and methodically, double-checking everything as you proceed, is the best way to ensure this, but you should also test your work when it is completed.

THINGS YOU NEED
- Continuity tester
- Socket outlet tester
- Qualified electrician

Right: Using a plug-in socket tester

When a new electrical installation is completed, it is inspected and tested to ensure that it complies with the relevant statutory requirements – in other words, that it complies with the IEE Wiring Regulations. Electricity supply companies can refuse to connect a supply to a consumer if the system does not meet electrical safety requirements.

Assuming that your system met the requirements of the Regulations to start with and that any wiring work you have carried out followed the instructions given in this book, it is reasonable to assume that the system will still meet the Regulations' requirements. However, there is nothing to be lost and considerable peace of mind to be gained by having the system tested by a qualified electrician; this is essential if you have carried out major changes such as adding extra circuits. Use an electrician who is a member of the Electrical Contractors' Association (ECA) or who is on the roll of the National Inspection Council for Electrical Installation Contracting (NICEIC). You can get lists of ECA and NICEIC electricians from your local electricity supply company's showrooms.

The test usually includes a visual inspection of the system, followed by:

- continuity of ring circuit conductors
- continuity of earth conductors and bonding links
- insulation resistance between live and neutral conductors and between live or neutral and earth conductors
- correct polarity (non-reversal of live and neutral connections)
- earth fault loop impedance
- earth electrode resistance (if applicable)
- operation of residual current devices.

Once the tests have been completed and assuming that no faults have been found, the electrician then issues a certificate stating that the system complies fully with the Wiring Regulations. This will probably include a recommendation that the system is inspected and tested at fixed intervals in the future too.

If you have carried out little more than the addition of a new socket outlet or the fitting of an extra light or two, a visual inspection of your work is generally all that is required to ensure that conductors are properly connected to the correct terminals. You can use a continuity tester to check the wiring on extensions to existing circuits (with the power turned off), and there is a widely available proprietary plug-in tester that can be used to indicate whether socket outlets have been correctly wired up. This is used with the power on.

REWIRING A HOUSE

PLANNING A REWIRE
DOING THE WORK

PLANNING A REWIRE

If your home's wiring system is clearly old-fashioned and cannot deliver the service you expect of it, what you need is a complete rewire to modern standards. There is no reason why you should not do the job yourself if you are a competent and careful worker, since it is a highly labour-intensive project, and the money you save can be spent on buying better-quality fixtures and fittings than you might otherwise have been able to afford.

Once you have made your mind up that rewiring your home is essential, you must aim to design your new wiring system to give you the service you want, both now and in the foreseeable future. To do this you will need to draw up some house floor plans – one for downstairs and one for upstairs. They do not have to be detailed architectural plans, but should be drawn approximately to scale on squared paper, to help you to estimate the quantities of cable you will need for the various circuits as well as enabling you to fit everything in the right place when you come to do the installation work.

Once you have drawn out the plans, take copies of each one so you can use one set for the lighting circuits, the other for the power circuits, at each of three levels – ground-floor level for downstairs power circuits, floor level upstairs for downstairs lighting circuits and upstairs power circuits, and the loft for the upstairs lighting circuits. It is best to plan out the lighting and power circuits separately, and this also makes for clearer drawings when you begin to work out how the circuit cables will run, even though the two will merge once the actual electrical installation work gets under way.

ASSESSING YOUR NEEDS

The first part of the planning process is to decide exactly what electrical services you will need in each room in the house. Start with the lighting circuits. The provision of a single pendant or other light fitting in the centre of each room is the absolute minimum requirement, but it is by no means essential if you have more ambitious plans for your lighting schemes, and can be dispensed with altogether if you intend to provide lighting by other means – wall lights, perimeter lighting, fittings recessed into the ceiling or concealed above worktops, and so on.

Using plug-in luminaire support couplers (LSCs) means you can have far greater flexibility in planning your lighting. With LSCs, you can provide lighting sockets on walls and ceilings at various points in every room, even if they are not all needed at present; once fitted, they can simply be blanked off with unobtrusive cover plates until they are.

As you work from room to room on your plans, mark the positions of individual lighting points and where you will want switches to control them. Remember that you can control an individual light from several linked switches if you want. Deal with 'utility' areas such as lofts and walk-in cup-

boards as well as the main rooms. You can also think about what you want in the way of outside lighting, both on the house walls and remote from the building, and whether you want lights in outbuildings.

Repeat the exercise on the other set of plans to pin-point the positions of socket outlets and fused connection units to serve individual appliances, and also the connection points for heavy current users such as the cooker, the immersion heater, an instantaneous shower or night storage heaters. Remember that on modern power circuits there is no restriction on the number of socket outlets each circuit can serve, only on the floor area of the rooms supplied by the circuit, so you may as well have too many as too few.

Always specify double socket outlets unless there is an overriding reason why a single one should be installed at a particular location. It may even be worth specifying triple outlets in rooms such as the kitchen and lounge where you are likely to want to use a lot of individual appliances. It is also worth thinking about the level at which the socket outlets will be installed; they are traditionally placed just above the skirting boards, but there are many situations where waist-level socket outlets would be much more practical. Mark these on your plan.

In the kitchen you will want most of the socket outlets fitted above worktop level, but do not forget to make provision for power supplies at a lower level to built-in (or parked-in) appliances such as washing machines, dishwashers, fridges and waste disposal units. Such outlets are best wired as spurs, each controlled by a double-pole switch sited above worktop level.

Lastly, think about whether you will want power supplies out of doors – either in outbuildings or down the garden. Once you have done this basic assessment of what electrical services you require, you can start to work out how this translates into individual circuits, and also begin to draw up your shopping list for the various wiring and lighting accessories you will need.

CIRCUIT PLANNING

Start with the lighting circuits. These are usually rated at 5 or 6 amps and so can in theory supply up to 12 lighting points, each of which is nominally rated at 100 watts. In practice it is best to restrict each circuit to a maximum of eight lighting points, to allow some of them to be fitted with high-wattage lamps (or several smaller ones). Since your living room is likely to be the one with the highest lighting demand, it is probably a good idea to allocate that its own circuit. Then, unless the rest of the house also has high demands, one circuit for

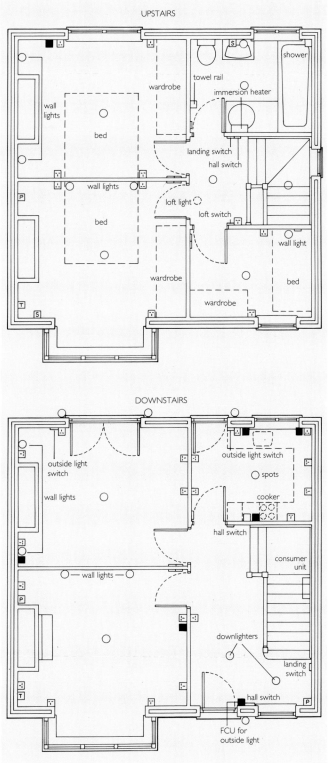

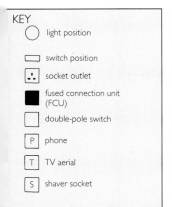

KEY

◯ light position

▭ switch position

⊡ socket outlet

◼ fused connection unit (FCU)

▢ double-pole switch

P phone

T TV aerial

S shaver socket

Left: Start with a plan of the outlets and fittings you require. Show lights and switches on a different set of plans from the socket outlets if this is easier. Add in fused connection units and other types of outlet, plus any non-mains wiring outlets. Include large current users which will need their own circuits

upstairs lights and another for the rest of the ground floor will probably be adequate. The occasional light on the outside wall of the house can be wired up to one of the indoor circuits as long as the circuit will disconnect within 0.4 seconds in the event of a fault, but extensive outdoor lighting needs its own separate circuit (essential for lights remote from the house).

Plan the circuits serving the socket outlets next. Most homes are wired up with ring main circuits, and a typical home will need just two circuits – one for upstairs, one for downstairs, or one for the front and one for the rear of the house. However, with the multitude of appliances in use in today's homes – especially the high current users in the kitchen such as washing machines, tumble driers and dishwashers – there could be a real risk of overloading the downstairs circuit. It therefore makes sense to allocate one ring circuit to the kitchen, one to other downstairs rooms and one to upstairs. Only in very large homes will additional circuits be needed to cope with the large floor area.

If the kitchen (or any other part of the house, for that matter) is remote from the incoming supply, ring circuits can be wasteful of cable. If the maximum load of the circuit is limited to 20 amps it would be cheaper to use a radial circuit where the cable terminates at the most remote socket outlet instead of returning to the consumer unit to complete the ring. However, the extra cost of the thicker cable required for a 32-amp radial circuit outweighs the financial advantage. You will also require radial circuits to outbuildings and to outdoor socket outlets.

You will now know how many lighting and power circuits your house will need. All that remains is to add the individual circuits needed for the cooker (provide this even if you plan to cook by gas, in case a future occupant of your home prefers to cook with electricity), the immersion heater, an electric shower and any night storage heaters, to arrive at the total number of separate circuits required. You can then choose a consumer unit with sufficient fuseways, ideally with a couple of additional spare fuseways to allow for any future extension to the wiring to be made with the minimum of disruption to the system.

Work out the total load of the new system and inform your electricity supply company to make sure that the cut-out fuse is large enough.

THE CONSUMER UNIT

When choosing the new consumer unit, be sure to specify miniature circuit breakers (MCBs) rather than rewirable or cartridge fuses to provide individual circuit protection. They are much more convenient than fuses, even if they are more expensive. Provide protection for vulnerable circuits against the risks of shock or fire caused by electrical faults by adding a residual current device (RCD) within the consumer unit. Circuits to outbuildings, or to outdoor lights and socket outlets, should also have high-sensitivity RCD protection.

You now have a detailed specification of what electrical equipment you want and where you want it. Use the checklist here to add up your requirements for the various wiring accessories that will be required for the job.

CIRCUIT CABLE RUNS

The next step is to think about how the wiring will actually be installed round the house. The most important departure from tradition that you can consider is how you actually run the circuit cables.

In these days of fitted carpets and built-in furniture, it is very inconvenient to run cables under floorboards, or to bury them in walls, where they are impossible to get at in the future without doing serious damage to the room's fittings and decorations. Modern cable trunking systems mean that much traditionally concealed wiring can instead be placed in neat, unobtrusive mouldings run round the room. However moulded trunking which replaces skirting, architraves and cornices is expensive, whereas installing cables in voids or buried in plaster is time consuming but costs little in terms of materials.

That just leaves the question of getting cables to wiring accessories mounted on walls or ceilings away from wall perimeters. Here you will have to cut wall chases and lift floorboards unless you are prepared to have surface-mounted mini-trunking, which is more obtrusive in these locations. In chases use conduit big enough to allow extra cables to be pulled through in the future.

Draw in the likely positions of cable runs for all the circuits on your plans, and use the scale selected to calculate approximately how much cable you will need. Add the figures to your parts checklist. It makes sense to buy the cable for lighting and power circuits in 50m or 100m drums since you will be using a lot of both, but buy larger cable sizes by the metre.

WIRING ACCESSORIES CHECKLIST

ITEM	No.	NOTES	ITEM	No.	NOTES
POWER CIRCUITS			4-terminal junction boxes		
Single socket outlets			Special light fittings		
Double socket outlets			OTHER CIRCUITS		
Triple socket outlets			Telephone socket outlets		
Metal-clad socket outlets			TV socket outlets		
2-amp round-hole socket outlets			TV/FM socket outlets		
Weatherproof outdoor sockets			Mounting boxes		
Unswitched fused connection units (FCUs)			Door bell and push		
Switched FCUs			Central heating controls		
Flex outlet plates			Burglar alarm		
20-amp double-pole (DP) switches			Smoke alarms		
20-amp DP switches with flex outlet			CONSUMER UNIT		
20-amp dual immersion heater switch			Consumer unit (no. of ways?)		
Immersion heater timer			5/6-amp MCBs		
30/32-amp DP switches			15/16-amp MCBs		
40/45-amp DP switches			20-amp MCBs		
Cord-operated DP ceiling switches			30/32-amp MCBs		
Cooker control unit			40/45-amp MCBs		
Cooker connection unit			Main isolator switch		
Shaver supply unit			Integral RCD		
Shaver socket outlet			Bell transformer		
Single flush mounting boxes			CABLE AND FLEX		
Single surface mounting boxes			1.0mm^2 two-core-and-earth cable		
Double flush mounting boxes			1.0mm^2 three-core-and-earth cable		
Double surface mounting boxes			2.5mm^2 two-core-and-earth cable		
Triple boxes			4.0mm^2 two-core-and-earth cable		
Special mounting boxes			6.0mm^2 two-core-and-earth cable		
30-amp junction boxes			10.0mm^2 two-core-and-earth cable		
2-amp plugs			0.5mm^2 two-core-and-earth flex		
LIGHTING CIRCUITS			0.5mm^2 two-core flex		
One-gang one-way plateswitches			1.0mm^2 two-core-and-earth flex		
One-gang two-way plateswitches			1.5mm^2 two-core-and-earth flex		
Two-gang two-way plateswitches			2.5mm^2 two-core-and-earth flex		
Multi-gang plateswitches			Heat-resisting flex		
Architrave plateswitches			Single-core meter tails		
Cord-operated ceiling switches			Single-core earth cable for cross-bonding		
One-gang dimmer switches			Earth clamps		
Two-gang dimmer switches			Coaxial aerial cable		
Metal-clad plateswitches			Telephone cable		
Weatherproof outdoor switches			Bell wire		
Flush mounting boxes			Multi-core flex for heating controls		
Surface mounting boxes			SUNDRIES		
Ceiling roses			PVC conduit and fittings		
Pendant lampholders			Mini-trunking and fittings		
Battenholders			Cable clips		
Luminaire support couplers			5- and 30-amp connector blocks		
Conduit (BESA) boxes			PVC insulating tape		
3-terminal junction boxes			Green/yellow PVC earth sleeving		
			Rubber grommets for knockouts		
			Plug fuses		

DOING THE WORK

Unless you are working in an empty house, your main problem throughout the job will be to minimise the disruption both to the existing electricity supplies and to everyday life. It is therefore best to split the job into manageable stages.

For example, if you have a two-storey house, divide it into three stages – ground-floor level (downstairs power circuits), first-floor level (downstairs lighting and upstairs power circuits) and loft level for the upstairs lighting circuits. There will be cross-over points – for example, where a downstairs switch controls an upstairs light, or a where a spur cable feeds a remote socket outlet – and each circuit cable must ultimately find its way down to the position of the new consumer unit, but these can be dealt with at the appropriate time as the project nears completion.

Throughout the job, the existing wiring is left undisturbed and a completely new system is fitted alongside it. When the new system is in place and is connected to the new consumer unit, the old wiring accessories and their mounting boxes are disconnected and removed, the old cables are cut back and abandoned and the recesses in the walls are plastered over ready for redecoration.

TOOLS AND MATERIALS

Before you start work, check that you have all the wiring accessories and cable you will need for each stage of the project; the checklists and plans you prepared earlier will be invaluable here. Do not forget sundries such as green/yellow PVC sleeving for bare earth cores, rubber grommets for lining the cable entry holes in metal flush mounting boxes, cable clips and connector blocks, and some red PVC insulating tape for flagging live cores in switches. You should also check that your toolkit contains all the necessary electrical and general tools.

WIRING DOWNSTAIRS

Where you start work is a matter of personal choice. Assuming that you have decided to begin with the wiring at ground-floor level, step one is to mark the positions of all the socket outlets and other power circuit wiring accessories on the room walls, and also the directions of the cable runs to and from each one.

If you plan to flush-mount all the accessories and you have solid walls, cut out the recesses for the mounting boxes next. Then cut chases from each accessory position down to floor level, ready for the cables to be run in, and fit the mounting boxes.

If you have plasterboard walls, use a padsaw to cut holes into which hollow wall boxes can be fitted later on, when the cables have been run in. It is best to avoid flush-mounting accessories on external walls if you have a timber-framed house, since you will risk damaging the vapour barrier lining the wall as you cut the holes.

If you prefer to surface-mount the accessories – permanently, or until you next redecorate the room and can flush-fit them then – simply secure the mounting boxes to the wall surfaces.

RUNNING IN THE CABLES

You can now start installing the cable runs that will link the various accessories to the consumer unit. How you do this will depend on whether you have concrete or timber ground floors, and if the latter, on whether you are willing or able to lift floorboards to gain access to the underfloor void.

If you can lift floorboards, you need to plan the precise cable directions with care to minimise the number you have to lift. First find out which way the floor joists run; it is easy to fish a cable through the void beneath the floor in any direction, but the cables should either pass through holes drilled in the joists or else be clipped to their sides rather than simply being left to dangle in the void. If there is enough space below the floor you can lift enough boards to allow you to climb down, and then clip all the cables in place. Otherwise you will have to lift say every fifth or sixth board so you can clip the cables to the joists. For cable runs across the lines of the joists, you need to lift just one board; then you can drill holes through the joist centres and thread the cable through the holes.

When you reach the wall beneath an accessory position, chop out the plaster behind the skirting board so you can feed the cable up into the wall chase you cut earlier and take it on up to the mounting box through conduit. Feed it into the box, leaving about 150mm (6in) protruding. If you are wiring a ring-main socket, feed the return cable back down the conduit and on to the next socket. For spurs fed from junction boxes, locate the boxes on joist sides at appropriate points and then run the spur cable in as already described. With plasterboard walls, the neatest way of getting cable to accessory posi-

tions is to prise off the skirting board and run the cable along the base of the wall to just below the mounting box in the gap between the bottom edge of the plasterboard cladding and the floor. Before replacing the board, make a small hole in the plasterboard just above floor level through which the cable(s) can be passed into the hollow centre of the wall and up to the box position. Then replace the skirting board.

With cable run to and from all the sockets, you can complete each ring circuit by running cable back to the consumer unit position from the first and last sockets on the ring. Leave enough cable at the consumer unit to facilitate connection later on. Radial circuits only have a cable running back from the first socket on the circuit to the consumer

unit. Follow the identical procedure to run cable from the consumer unit position for individual circuits to appliances such as the cooker and any night storage heaters, using cable of the appropriate size. If you have solid floors, or are not prepared to disturb fitted floorcoverings and furniture, the best solution is to place all the floor-level cable runs in trunking of some sort fixed at skirting-board level. The neatest is skirting-board trunking, which fits over the existing skirting board and conceals the cables run along its top edge. Chases in the plaster then run up from the trunking to the accessory mounting boxes. With all the circuit cables in place for this stage of the work, you can then start fitting the new wiring accessories.

As with the first stage of the job, remember that the object of the exercise is to install the new wiring system alongside the existing one where possible so as not to disrupt everyday life too much. The old wiring will simply be disconnected and abandoned when the new cables and wiring accessories are in place; the old ones will then be removed and their mounting positions plastered over.

UPSTAIRS POWER CIRCUITS

Begin by marking the positions of all the new socket outlets in the upstairs rooms, and then map out the best routes for the upstairs power circuit cables to take. Remember that you can use spurs to reach individual socket positions that do not fit conveniently on the ring circuit route, and you can also drop spurs down to ground-floor

level if you wish to reach parts of the house remote from the downstairs ring. The ends of the ring circuit cable will have to drop down to the consumer unit position at a convenient point; remember to leave ample cable there for later connection to the new consumer unit.

Next, chop out all the cable chases and mounting-box recesses as for the downstairs power circuits. If you plan to use skirting trunking for wiring up the power circuits, you can simply run cables round rooms at skirting-board level; you will only have to lift a floorboard to take the circuit cables downstairs back to the consumer unit position.

If you prefer a conventional underfloor installation, check which way floorboards run in each room, so you can decide on the most convenient route for the cables to take. If you

WIRING UPSTAIRS

With the rewiring at ground level as complete as possible, it is now time to start work on the upstairs power circuits and the downstairs lighting circuits. The wiring for both will be run mainly in the ground-floor ceiling void.

have to run cables parallel to the joist direction, lift a board at each side of the room and check with a mirror and torch to see whether the void is clear of obstructions. If it is and the ceiling beneath is of plasterboard, you should be able to push the cable through from one side of the room to the other; otherwise you will have to lift one or more intermediate boards and fish the cable through the void with a length of stiff wire.

Feed the cable into each outlet recess, working round the ring until all have been supplied. Take the cable drop(s) down to the consumer unit position. With the cable runs complete, position and secure all the mounting boxes, feed the cable into each one and connect the outlets.

BATHROOM WIRING

Remember that you may not install socket outlets in bathrooms, except for special shaver sockets containing a transformer which can be wired directly from the upstairs or downstairs lighting circuit or via a fused connection unit on the upstairs power circuit.

To provide power for electric wall heaters and towel rails in the bathroom, use fused spurs from the upstairs ring circuit. For a wall heater with its own switch, simply run cable from the FCU directly to the heater. For towel rails and heaters without a switch, run cable from the connection unit to a ceiling-mounted cord-operated switch in the bathroom or a switch outside the room, then on to a flex outlet plate next to the heater. Link the heater to the flex outlet plate with three-core flex.

OTHER UPSTAIRS CIRCUITS

If you have an electric immersion heater or an electric shower, wire their new individual radial circuits up next. Run the cable from the consumer unit to the control switch position, and mount a double-pole isolating switch there; remember to position the switch out of reach of any-

one using the bath or shower. Connect in the circuit cable, then link the appliance to the switch. Cross-bond the shower supply pipework to earth along with all the other bathroom metalwork.

DOWNSTAIRS LIGHTING CIRCUITS

You can now turn your attention to the wiring for the downstairs lighting circuits. Start by lifting upstairs floorboards (or cutting out board sections) above the positions of all downstairs ceiling-mounted light fittings so you can disconnect the old cables and remove the old fittings. Then mount the new light fittings in position, ready for the new circuit cables to be connected to them. Screw ceiling roses to joists where possible. For other fittings or where joists do not coincide with the desired rose position, fix battens on blocks between the joists so the rose or light can be screwed through the ceiling surface into the batten above.

Most of your wiring is likely to be of the loop-in type, but you can mix in elements of junction-box wiring if this makes for more economical use of cable. However, use of junction-box wiring does mean that the circuit connections for light fittings are inaccessible without lifting floorboards. As with the power circuit wiring described earlier, run the circuit cables through holes drilled in the joists where they cross the joist line; leave them resting on the ceiling surface where they run parallel to the joists. Finally, take the cable drop from the first light on the

circuit back down to the consumer unit position, ready for connection later.

Next, decide where you want the downstairs light switches to be positioned, chop out recesses for the new mounting boxes and cut chases down to them from ceiling level. If you are lucky enough to find that the old switch drops have been run in conduit and you want the new switches in the same places as the old ones, disconnect the old cables and use them to draw the new cable into the conduit. You will have to switch off the existing wiring system at this point. Otherwise simply abandon the old switch cables once the new wiring is in place. Now you can fit the new light switches and connect up the switch drops.

The stairwell is one area where the upstairs and downstairs lighting meet, since you will probably want two-way switching of the hall and landing lights from both floors. Both lights must be powered from only one of the circuits, so make provision for this when planning out the circuits. The switches are linked by two three-core-and-earth cables.

If you are installing wall lights in downstairs rooms, cut chases in the walls up to ceiling level so you can run in their supply cables from the ceiling void above. Where one switch will control two or more wall lights, link their supply cables at a three-terminal junction box and run one cable from there to a four-terminal box on the lighting circuit. Wire the switch cable into this box.

Installing the wiring for the upstairs lighting circuit is the easiest part of the whole job, so long as you have a loft and it is not packed to the rafters with bric-a-brac, since you can simply clip the cables to the ceiling joists and gain easy access to the positions of individual lighting points in the rooms below.

If your house has a flat roof, you may be able to fish some new cables across the ceiling void, but it is generally easier to surface-mount them in unobtrusive mini-trunking unless you are prepared to take down and replace the ceilings.

Assuming that you do have a loft, start by disconnecting the old supply cables to all the upstairs lighting points, and mount any new light fittings in position, ready for the new circuit cables to be connected to them. Screw ceiling roses to joists where possible. For other fittings or where joists do not coincide with the desired rose position, fit battens on blocks between the joists to provide a secure fixing point.

As with the downstairs lighting circuit, most of your wiring is likely to be of the loop-in type, but you can use elements of junction-box wiring if this makes for more economical use of cable. Where cable runs are parallel with the joists, simply clip the cable in place to the upper side face of each joist, above any loft insulation. If you have to cross the joist lines, take them over each joist at a point where they cannot be stepped on.

Do not run them through holes drilled in the centres of the joists, as elsewhere on the system, because this would effectively enclose the cable in the loft insulation. While you are working in the loft, take

the opportunity to fit one or more lights there if you do not have one already. Wire them up as an extension to the upstairs light circuit, and position the switch that will control them somewhere convenient on the floor below, ideally close to the position of the loft hatch.

Finally, take the cable drop from the first light on the circuit back down to the consumer unit position, ready for final connection later on. If you are installing wall lights in upstairs rooms, cut chases in the walls up to ceiling level so you can run in their supply cables from the loft above. Where one switch will control two or more wall lights, link their supply cables at a three-terminal junction box and run one cable from there to a four-terminal box on the lighting circuit. Wire the switch cable into this box to provide the switching.

BATHROOM WIRING

Any light fittings installed in bathrooms must either be fully enclosed or must have lamp-holders with extra-deep skirts so there is no risk of fingers touching live metal as the lamp is changed. Use a ceiling-mounted cord-operated switch unless the room is big enough for the switch to be sited out of reach of anyone using the bath or shower – more than 2.5m (8ft 3in) away.

Special shaver sockets containing a transformer can be wired as a spur from the upstairs or downstairs lighting circuit or from the upstairs power circuit, whichever is the more convenient. This is also a good opportunity to provide wiring for an extractor fan, which will help reduce condensation problems in the room. This can be wired as a

spur from the lighting circuit, and in windowless rooms it can be controlled by the light switch so it is always switched on whenever the room is in use.

THE CONSUMER UNIT

With all the new circuit wiring complete and all the new wiring accessories and lights connected up, it is time to install the new consumer unit and make the final circuit connections. Mount the new unit on a flameproof backing board next to the existing fusebox (unless you want it sited somewhere more convenient: if you do, ask your local electricity board if they can extend the main incoming supply cable to a new board position).

Then select and mount the circuit fuses or circuit breakers you need to match the various circuits you have installed. Fit the fuseways in descending order of current rating, with the highest-rated next to the main isolating switch or residual current device, unless you are fitting a split-load unit.

Next, feed in each circuit cable in turn, cutting off excess cable as necessary after making sure the cable cores will reach the neutral and earth terminal blocks as well as the live connection at the fuseway itself. Strip off the outer

THE FINAL STAGES

When the rewiring work on your new power circuits and the downstairs lighting circuit is complete, you can move up to the loft and start work on the upstairs lighting circuits. Then you can finish off the job by connecting all the circuit cables up to your new consumer unit, and removing the remains of the old wiring installation.

Below: Downstairs socket outlets and lighting circuits run in the ceiling void.

sheathing and prepare each core in turn, arranging them neatly in the box as you work along the row of fuseways. Label each fuseway for identification.

At this stage, complete the cross-bonding to earth of all exposed metalwork – baths, showers, kitchen sinks and exposed plumbing, central heating and gas supply pipework – using special earth clamps and single-core earth cable. Run the earth cable back to the consumer unit position and connect it to the main earthing terminal there.

THE FINAL CHANGEOVER

Now all that remains is to call in a professional electrician or your local electricity board to test the installation (especially for correct polarity, earthing effectiveness and insulation resistance), issue a test certifi-

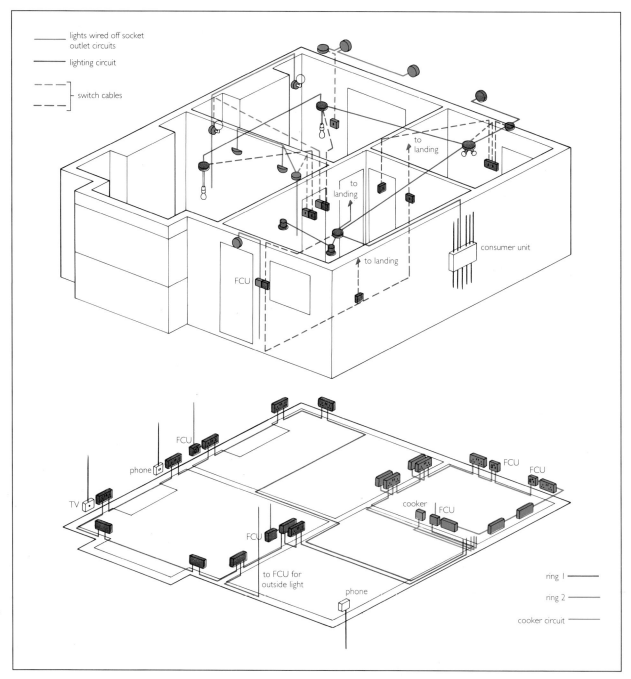

lights wired off socket outlet circuits

lighting circuit

switch cables

to landing

to landing

to landing

consumer unit

FCU

FCU

phone

TV

FCU

cooker

FCU

FCU

FCU

to FCU for outside light

phone

ring 1

ring 2

cooker circuit

cate and switch the main supply and earth connections over from the old fusebox to the new consumer unit. You cannot do this yourself. Once that has been done, you can then disconnect and remove the old fusebox and any remaining old wiring accessories and mount-

ing boxes, make good their recesses and remove or abandon the old cables leading to them.

Below: Upstairs socket outlets and lighting circuits run in the loft. The diagrams show the individual circuits and cable runs required to

supply the system plan shown on page 145.

Note that the landing lights are powered by the downstairs lighting circuit. Wall lights and those on the outside walls are supplied in a variety of ways to make efficient use of cable and to avoid overloading lighting circuits.

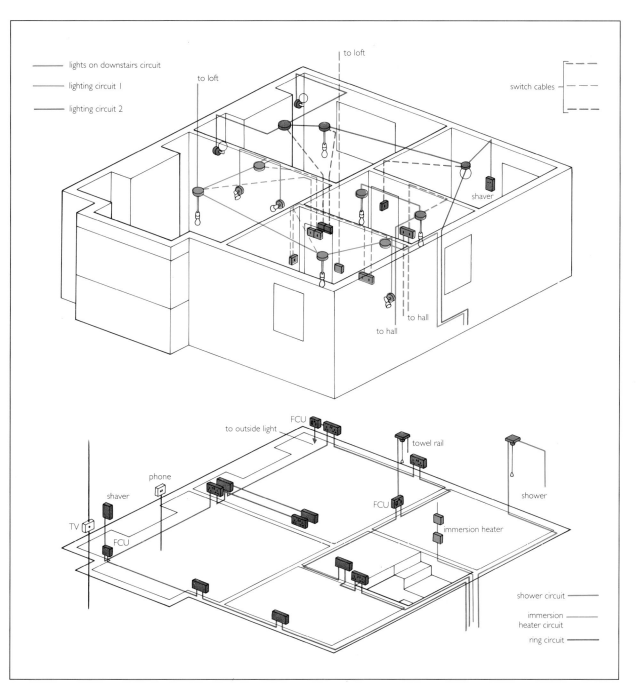

153

GLOSSARY

AMPERE
The unit of electrical current, often abbreviated to amp or A. To work out how much current an appliance uses divide its wattage by the mains voltage (240V).

ARCHITRAVE SWITCH
A narrow plateswitch, designed for mounting on door architraves or in other locations where space is limited, available in one- and two-gang version.

BATTENHOLDER
A utility light fitting for wall or ceiling mounting, consisting of a straight or angled lampholder attached directly to a circular baseplate.

BUSBAR
A solid conducting bar in a consumer unit, on which the individual circuit fuses or MCBs are mounted.

CABLE
Used for wiring up the circuits making up a house wiring system, cable has one or more metallic insulated conductors (or cores) covered with a protective outer sheath. If the cable contains an earth core, this is uninsulated within the cable sheath.

CARTRIDGE FUSE
A fuse consisting of fuse wire within a tubular holder, fitted in some older-style fuseboxes and in modern 13-amp plugs and fused connection units.

CATENARY WIRE
Galvanized steel wire, which must be independently earthed, used to support overhead cable runs between buildings.

CEILING ROSE
A round wiring accessory that provides a permanent connection between the circuit wiring and the flex to a pendant lampholder. *See also* Luminaire support coupler.

CIRCUITS
Complete paths round which current flows – along the live conductor to where it is needed, then back to its source along the neutral conductor.

CONDUIT
Round, oval or square PVC tube, used to protect cable runs beneath plaster, out of doors or underground.

CONDUIT BOX
A plastic or metal mounting box usually circular in shape, also called a BESA box, mainly used to contain the electrical connections to some wall and ceiling light fittings.

CONSUMER UNIT
The main control box that governs the distribution of electricity to all circuits in the house. The unit contains the system's main on-off switch, and fuses or circuit breakers protecting each individual circuit in the house.

COOKER CONTROL UNIT
A double-pole switch to control a built-in or free-standing cooker, combined with a switched socket outlet.

CORE
Any one of the current-carrying or earth conductors within flex and cable.

CPC (CIRCUIT PROTECTIVE CONDUCTOR)
See Earthing, Protective conductor.

CROSS-BONDING
The linking of exposed metalwork – water pipes, sinks, baths, towel rails and so on – to each other and on to earth for safety purposes.

DIMMER SWITCH
A plateswitch that allows the brightness of the light it controls to be varied.

DISTRIBUTION BOX
A special heavy-duty terminal box used to split the incoming mains supply to feed two or more separate consumer units. Also known as a Henley box or splitter box.

DIVERSITY
A method of calculating the likely current demand of a circuit, taking into account the fact that it will not be supplying the maximum theoretical demand at any time.

DOUBLE-POLE (DP) SWITCH
An on-off switch that cuts both the live and neutral sides of the circuit, so completely isolating the appliance it controls from the mains supply when it is switched off.

EARTHING
The provision of a continuous conductor on circuits to help protect the user from the consequences of certain electrical faults. Earth conductors are insulated with green/yellow striped PVC in flex, and are covered with slip-on green/yellow PVC sleeving where they are exposed in cables.

EARTH LEAKAGE CIRCUIT BREAKER
The former name for a protective device now referred to as a residual current circuit breaker (RCCB) or residual current device (RCD).

EQUIPOTENTIAL BONDING

The use of protective conductors to interconnect all exposed conductive parts and extraneous accessible metalwork that could become live under fault conditions, even if the latter does not form part of the electrical installation.

FLEX

This is an abbreviation from flexible cord, and consists of insulated conductors within a flexible outer sheath. It is used to link appliances and pendant lights to the mains supply.

FLEX OUTLET PLATE

A wiring accessory that connects an appliance flex permanently to the circuit wiring.

FUSEBOX

An old-style control unit containing the system's main on-off switch and the fuses protecting individual circuits in the house. It has now been superseded by the modern consumer unit.

FUSED CONNECTION UNIT (FCU)

A wiring accessory that connects an appliance permanently to the mains supply, instead of it being plugged in at a socket outlet. It can also supply a fused spur.

FUSED SPUR

A branch line off a main power circuit, protected by a fuse of lower rating than the circuit fuse.

FUSES

Protective devices inserted into electrical circuits and fused plugs to provide short-circuit protection and to prevent overloading.

GANGS

A term that describes the number of individual switches or socket outlets contained in one wiring accessory.

GROMMETS

Small plastic washers used in metal mounting boxes to stop the cable sheathing from chafing on the edges of a knockout.

HENLEY BOX

See Distribution box.

IEE WIRING REGULATIONS

A document that lays down guidelines for safe electrical installation practice, issued by the Institution of Electrical Engineers. The Regulations do not have legal force in England and Wales, but in Scotland they form part of the Building Regulations, and so are enforceable. The current edition (the 16th) is now a British Standard – BS7671:1992. Its requirements are followed throughout this book.

INSULATION

A plastic covering on the cores of flex and cable that protects users of electrical equipment and appliances from touching live conductors.

JUNCTION BOX

An accessory used on power circuits to connect in spurs, and on lighting circuits wired on the junction-box principle to link the circuit cables to each ceiling rose or light fitting and its switch. They are also known as joint boxes.

KNOCKOUT

A pre-formed weak spot in metal and plastic mounting boxes, designed to be knocked out as needed to admit the circuit cables.

LAMP

The 'trade' term for a light bulb or tube.

LAMPHOLDER

An insulated metallic socket into which lamps are plugged. Pendant types are linked to their ceiling roses by flex.

LIVE

A term that describes the cable or flex core carrying current to a wiring accessory or appliance, or any terminal to which this core is connected. The core insulation is colour-coded for identification: the live core is red in cables, brown in flex.

LOOP-IN LIGHTING CIRCUIT

A circuit wired by running cable to each ceiling rose or light fitting in turn. The switch cable is wired directly to the rose or fitting it controls.

LUMINAIRE SUPPORT COUPLER (LSC)

A plug-and-socket system used for pendant ceiling lights and wall lights, allowing the light fitting to be unplugged for cleaning or maintenance.

MINIATURE CIRCUIT BREAKER (MCB)

An automatic switch fitted to each circuit in the consumer unit which switches off in the event of a short circuit or an overload on the circuit.

MINI-TRUNKING

Surface-mounted PVC channelling with a U-shaped body and a snap-on cover, designed to conceal circuit cables and to protect them from damage.

MOUNTING BOX

A square or rectangular box in metal or plastic which can be flush or surface-mounted. The accessory faceplate is screwed to lugs at each side of the box.

NEUTRAL

A term that describes the cable or flex core carrying current back to its source, or any terminal to which this core is connected. The neutral conductor is a live conductor: the term does not imply that it is dead. The core insulation is colour-coded for identification: the neutral core is black in cable, blue in flex.

ONE-WAY SWITCH

A plateswitch that controls a light from one switch position only. It contains just two terminals per gang, often marked L1 and L2.

PATTRESS

The backplate of fittings such as ceiling roses and cord-operated switches.

PHASE CONDUCTOR

The live conductor in a wiring system, describing any conductor not connected to the neutral point of the system nor acting as a protective conductor.

PLATESWITCH

The 'trade' term for a wall-mounted light switch.

POLARITY

Correct polarity means that live cores go only to live terminals and neutral cores only to neutral ones; if they are wired the other way round, reverse polarity results.

PROTECTIVE CONDUCTOR

A conductor used to connect the various components of a wiring system to earth to provide protection against electric shock.

RADIAL CIRCUIT

A power circuit originating at the consumer unit and terminating at the most remote socket outlet, or at an individual electrical appliance.

RCBO

The (very necessary) abbreviation for residual current circuit breaker with overcurrent protection – a combined RCD and MCB.

RESIDUAL CURRENT DEVICE (RCD)

A protective safety device fitted to circuits to detect current leakage which could start a fire or cause an electric shock. Now widely used to protect users of power tools and electrical appliances out of doors.

RING CIRCUIT

A power circuit wired as a continuous loop, both ends being connected to the same terminals in the consumer unit.

SHAVER SOCKET OUTLET

A special socket outlet designed to supply electric shavers, made to BS4573, designed for use in rooms other than bathrooms and washrooms.

SHAVER SUPPLY UNIT

A socket outlet for electric shavers containing an isolating transformer, made to BS3535 and suitable for use in bathrooms and washrooms.

SINGLE-POLE (SP) SWITCH

A switch that cuts only the live side of the circuit it controls. Most plateswitches are of this type.

SOCKET OUTLET

A wiring accessory with three recessed terminals into which the pins of three-pin plugs fit, allowing appliances to be connected to and disconnected from the mains.

SPLITTER BOX

See Distribution box.

SPUR

A cable 'branch line' connected to a wiring circuit to supply extra lights or a socket outlet.

STRIP CONNECTOR

A metal barrel terminal encased in plastic with two terminal screws, used to link cable to cable or to flex. They are manufactured in strips, hence the name.

SWITCHFUSE UNIT

A unit containing just one fuseway, used to supply an extra circuit and fitted alongside the existing fusebox or consumer unit.

TWO-WAY SWITCHES

Switches that are used in pairs to allow control of one light from two switch positions. Each gang has three terminals, allowing the switches to be linked by special three-core-and-earth cable.

UNIT

A measure of the amount of electricity consumed by an appliance or circuit, used for pricing purposes by electricity supply companies. It is the product of the power consumed (measured in watts) and the time during which it was supplied. One unit is 1 kilowatt-hour (kWh) – the consumption of a 100-watt bulb burning for 10 hours, for example.

VOLT

The unit of electrical 'pressure' – the potential difference that drives current round a circuit. Usually abbreviated to V. In most British homes mains voltage is 240V. The term 'extra-low voltage' means a voltage of less than 50 volts.

WATT

The unit of electrical power consumed by an appliance, usually abbreviated to W. It is the product of mains voltage x current drawn (in amps). 1,000W = 1 kilowatt.

USEFUL ADDRESSES

MANUFACTURERS

Ashley & Rock Ltd,
Morecambe Road, Ulverston,
Cumbria LA12 9BN
Tel: (0229) 583333
Wiring accessories

B&R Electric Ltd,
Temple Fields, Harlow,
Essex CM20 2BG
Tel: (0279) 434561
Wiring accessories

Contactum Ltd,
Victoria Works, Edgware Road,
London NW2 6LF
Tel: 081-452 6366
Wiring accessories

Crabtree Electrical Industries Ltd,
Lincoln Works, Walsall,
West Midlands WS1 2DN
Tel: (0922) 721202
Wiring accessories

Crompton Parkinson Ltd,
Wheatley Hall Road,
Wheatley,
Doncaster DN2 4NB
Tel: (0302) 321541
Lamps and lighting equipment

GE Thorn Lamps Ltd,
Miles Road, Mitcham,
Surrey CR4 3YX
Tel: 081-640 1221
Lamps and lighting equipment

Home Automation Ltd,
Bumpers Way, Chippenham,
Wiltshire SN14 6LF
Tel: (0249) 443515
Wiring accessories

Legrand Electric Ltd,
Foster Avenue,
Woodside Park,
Dunstable,
Bedfordshire LU5 5TA
Tel: (0582) 609261
Wiring accessories

Marbourn Ltd,
Oakesway Road,
Hartlepool Industrial Estate,
Hartlepool,
Cleveland TS24 0RE
Tel: (0429) 265511
Wiring accessories

MEM Ltd, Whitegate,
Broadway, Chadderton,
Oldham OL9 9QG
Tel: 061-652 1111
Wiring accessories

MK Electric Ltd,
Shrubbery Road,
Edmonton,
London N9 0PB
Tel: 081-803 3355
Wiring accessories

Osram Ltd,
PO Box 17,
East Lane,
Wembley,
Middlesex HA9 7PG
Tel: 081-904 4321
Lamps and lighting equipment

Ottermill Ltd,
Ottery St Mary,
Devon EX11 1AG
Tel: (0404) 812131
Wiring accessories

Philips Lighting Ltd,
City House,
420–430 London Road,
Croydon,
Surrey CR9 3QR
Tel: 081-665 6655
Lamps and lighting equipment

Ring Lighting (Ring Lamp) Co Ltd,
Gelderd Road,
Leeds,
West Yorkshire LS12 6NB
Tel: (0532) 791791
Lamps and lighting equipment

Smiths Industries Environmental
  Controls Co Ltd,
Aspley Way,
off Waterloo Road,
London NW2 7UR
Tel: 081-450 8944
Automatic controls

Superswitch Electric
  Appliances Ltd,
Houldsworth Street, Reddish,
Stockport, Cheshire SK5 6BP
Tel: 061-431 4885
Automatic controls

Volex Accessories,
Leigh Road,
Hindley Green,
Wigan,
Lancashire WN2 4XY
Tel: (0942) 57100
Wiring accessories

USEFUL ORGANISATIONS

Electrical Contractors Association
  (ECA),
34 Palace Court,
London W2 4HY
Tel: 071-229 1266

Institution of Electrical Engineers
  (IEE),
Savoy Place,
London WC2R 0BL
Tel: 071-240 1871

National Inspection Council for
  Electrical Installation Contracting
  (NICEIC),
Vintage House,
34 Albert Embankment,
London SE1 7UJ
Tel: 071-582 7746

INDEX